Sniper Training and Employment

Department of Defense

All rights reserved. No part of this publication may be reproduced, stored in a retrieval system, or transmitted in any form or by any means, electronic, mechanical, photocopying or otherwise, without the prior permission of the copyright owner.

© Copyright 2008 – BN Publishing

Cover Design: J. Neuman

www.bnpublishing.net

info@bnpublishing.net

ALL RIGHTS RESERVED

For information regarding special discounts for bulk purchases, please contact BN Publishing

sales@bnpublishing.net

*TC 23-14

TRAINING CIRCULAR   HEADQUARTERS
No. 23-14   DEPARTMENT OF THE ARMY
                                       Washington, DC, 14 June 1989

SNIPER TRAINING AND EMPLOYMENT

CONTENTS

                                                          Page

PREFACE   iv

CHAPTER 1.   INTRODUCTION

     1-1.   Historical Background   1-1
     1-2.   Mission   1-3
     1-3.   Organization   1-4
     1-4.   Personnel Selection   1-5

CHAPTER 2.   EQUIPMENT

Section I.   Sniper Rifles   2-1

     2-1.   M21 Sniper Weapon System   2-1
     2-2.   M24 Sniper Weapon System   2-5

Section II.   Sniperscopes   2-15

     2-3.   Auto-Ranging Telescope   2-15
     2-4.   Leupold M3A Telescope   2-24

Section III.   Ammunition   2-26

     2-5.   Special Ball   2-26
     2-6.   Blanks   2-27

Section IV.   Optical Observation Equipment   2-27

     2-7.   Observation Telescope   2-27
     2-8.   Binoculars   2-28
     2-9.   Night Vision Sight   2-30
     2-10.   Night Vision Goggles (AN/PVS-5)   2-31

DISTRIBUTION RESTRICTION: Approved for public release; distribution is unlimited.

*This publication supersedes TC 23-14, 27 October 1969.

i

		Page
Section V.	Clothing and Additional Equipment	2-32
2-11.	Camouflage	2-32
2-12.	Additional Equipment	2-33

CHAPTER 3. SNIPER MARKSMANSHIP

3-1.	Using the Fundamentals of Marksmanship..	3-1
3-2.	Following Through	3-9
3-3.	Calling the Shot	3-10
3-4.	Zeroing the Rifle	3-11
3-5.	Considering Weather Effects	3-12
3-6.	Firing One Round	3-17
3-7.	Holding Off for Elevation or Wind	3-18
3-8.	Engaging Moving Targets	3-19

CHAPTER 4. FIELD TECHNIQUES

4-1.	Camouflage	4-1
4-2.	Cover and Concealment	4-6
4-3.	Movement and Land Navigation	4-8
4-4.	Selection and Occupation of Sniper Positions	4-17
4-5.	Construction of Sniper Positions	4-20
4-6.	Observation and Target Selection	4-28
4-7.	Information Records	4-32
4-8.	Range Estimation	4-40

CHAPTER 5. EMPLOYMENT

5-1.	Sniper Teams	5-1
5-2.	Offensive Employment	5-2
5-3.	Defensive Employment	5-3
5-4.	Retrograde Employment	5-3
5-5.	MOUT Employment	5-4
5-6.	Countersniper Operations	5-4

APPENDIX A.	FIELD TRAINING EXERCISES	A-1
APPENDIX B.	SNIPER SUSTAINMENT PROGRAM	B-1
APPENDIX C.	SNIPER'S DATA CARD	C-1
APPENDIX D.	MEASUREMENTS	D-1
APPENDIX E.	REFERENCE TABLES	E-1
APPENDIX F.	SNIPER PATROL ORDERS	F-1

	Page
APPENDIX G. REPRODUCIBLE FORMS	G-1
GLOSSARY	Glossary-1
REFERENCES	References-1
INDEX	Index-1

TC 23-14

PREFACE

This circular provides doctrine for the tactical employment of the infantry sniper. It also provides the information needed to train and equip snipers and to plan their missions and operations. It is intended for use by commanders, staffs, instructors, and soldiers at training posts, Army schools, and units.

The proponent of this publication is HQ, TRADOC. Submit changes for improving this publication on DA Form 2028 (Recommended Changes to Publications and Blank Forms) and forward to the Commandant, US Army Infantry School, ATTN: ATSH-IN-2B, Fort Benning, GA 31905-5595.

Unless otherwise stated, whenever the masculine gender is used, both men and women are included.

TC 23-14

CHAPTER 1
INTRODUCTION

A sniper has special abilities, training, and equipment. His job is to deliver discriminatory, highly accurate rifle fire against enemy targets which, because of range, size, location, fleeting nature, or visibility, cannot be engaged successfully by the rifleman. Sniping requires the development of basic infantry skills to a high degree of perfection. A sniper's training incorporates a wide variety of subjects designed to increase his value as a force multiplier and ensure his survival on the battlefield. The art of sniping requires learning and repetitiously practicing these skills until mastered. A sniper must be highly trained in long-range rifle marksmanship and fieldcraft skills to ensure maximum probability of effective engagements and minimum risk of detection. Perfection must be reached before a sniper takes part in combat operations.

1-1. HISTORICAL BACKGROUND

The term "sniper" originated in the 19th century with the British Army in India where the snipe was a favorite game fowl. The snipe is small and fast, and an extremely difficult target. The successful snipe hunter was an expert shot and proficient in other arts of the hunter. Therefore, the term "sniper" came to signify one who possessed all the skills of a successful snipe hunter. However, the proficiency of the military sniper evolved into an art as advancements in weapons, equipment, and techniques were made.

    a. The use of sharpshooters (or snipers) can be traced in US military history from the Revolutionary War. During the American Civil War, General Hiram Berdan was an exponent of the art and helped perfect the techniques used by snipers.

    b. In World War I, the British Army encountered expert German marksmen equipped with special rifles and telescopic sights. The term "sniper" was applied and popularized. German snipers forced the British Army to employ the same techniques, and under the leadership of Major Hesketh-Pritchard, a sniper course (the first Army School of Sniping, Observing, and Scouting) was organized. By the end of the war, the British were able to beat the Germans at their own game.

    c. After World War I, the emphasis on sniping decreased -- except in Soviet Russia. In 1930, Russia began training and equipping snipers. By World War II, they had carefully integrated sniper tactics into their tactical doctrine so that their snipers could operate as a well-drilled team. Each man knew exactly where to move and what to do.

d. During World War II, the US Army armed unit marksman with an M1, M1C, or M1903 Springfield rifle to conduct sniper activity. The results and effects differed between commanders and units. A specific lesson learned in the employment of snipers was that a sniper is a weapon of opportunity -- a typical rifleman cannot be arbitrarily assigned the sniper mission. Every marksman is not a sniper, but every sniper is a marksman.

e. Combat in Korea, using US Army and Marine Corps units, again reflected a lack of command appreciation for the techniques of employment and capabilities of snipers. American units, equipped with a new sniper rifle (the M1D with M84 telescopic sight), seldom relied upon snipers, although countersniping and interdiction by sniper fire was used in some instances. Recommendations resulting from the Korean war included the need for centralized sniper schools, a flexible sniper organization, use of skilled personnel, and the need to train commanders how to use a sniper's capabilities correctly. As a result, the United States Army Infantry School was tasked with the mission of organizing a sniper school. This mission was undertaken in coordination with the United States Army Marksmanship Training Unit during 1955 and 1956. The program reiterated the lessons learned:

(1) The best active protection against enemy snipers is a trained sniper.

(2) The skills required of a trained sniper must be superior to the average rifleman.

(3) A sniper must be a skilled shooter with a specialized weapon.

(4) A sniper must be well-trained in the combat skills of the individual soldier.

(5) Unstructured, incomplete training and the lack of doctrine inhibit the use of snipers.

(6) Education of commanders is vital to ensure the proper use of a sniper.

This program was short-lived because of the lack of understanding and appreciation throughout the Army for the value of a sniper. With the adoption of the M14 service rifle, no provision was made for an M14 sniper rifle. The designation of a sniper in the rifle squad was discontinued. The sniper training program became optional.

   f. The conflict in Vietnam revived the need for snipers. Enemy forces in that conflict demonstrated the effectiveness of sniper employment techniques under varying tactical conditions. The US Army conducted division-level sniper training courses and educated commanders at all levels on the use of snipers.

   g. During operation "Urgent Fury" in 1983, the US Army Rangers employed snipers in Grenada. Target reductions were successful against enemy mortar positions at ranges up to 800 meters. The reduction of fires from these positions was critical to the mission's success and illustrates the continuing value of sniper employment.

1-2. MISSION

The primary mission of a sniper in combat is to support combat operations by delivering precise long-range fire on selected targets. By this, the sniper creates casualties among enemy troops, slows enemy movement, frightens enemy soldiers, lowers morale, and adds confusion to their operations. The secondary mission of the sniper is that of collecting and reporting battlefield information.

   a. A well-trained sniper, combined with the inherent accuracy of his rifle and ammunition, is a versatile supporting arm available to an infantry commander. The importance of the sniper cannot be measured simply by the number of casualties he inflicts upon the enemy. Realization of the sniper's presence instills fear in enemy troop elements and influences their decisions and actions. A sniper enhances a unit's firepower and augments the varied means for destruction and harassment of the enemy. Whether a sniper is organic or attached, he will provide that unit with extra supporting fire. The sniper's role is unique in that it is the sole means by which a unit can engage point targets at distances beyond the effective range of the service rifle. This role becomes more significant when the target is entrenched or positioned among civilians, or during riot control missions. The fires of automatic weapons in such operations can result in the wounding or killing of noncombatants.

b. Snipers are employed in all levels of conflict. This includes conventional offensive and defensive combat in which precision fire is delivered at long ranges. It also includes combat patrols, ambushes, countersniper operations, forward observation elements, military operations on urbanized terrain, and retrograde operations in which snipers are part of forces left in contact or as stay-behind forces. Chapter 5 discusses sniper employment techniques in detail.

1-3. ORGANIZATION

In light infantry divisions, the sniper element is comprised of six battalion scouts organized into three 2-man teams. They may perform dual missions, depending on the need. In the mechanized infantry battalions, the sniper element is comprised of two riflemen (one team) located at each rifle company headquarters. The commander designates missions and priorities of targets for the team and may attach or place the team under the operational control of a company or platoon. In some specialized units, snipers may be organized according to the needs of the tactical situation.

a. Sniper teams should be centrally controlled by the commander or the sniper employment officer. The SEO is responsible for the command and control of snipers assigned to the unit. In light infantry units, the SEO will be the scout platoon leader or the platoon sergeant. In heavy or mechanized units, the SEO will be the company commander or the executive officer. The duties and responsibilities of the SEO are:

(1) Advising the unit commander on the employment of snipers.

(2) Issuing orders to the team leaders.

(3) Assigning missions and types of employment.

(4) Coordinating between the sniper team and unit commander.

(5) Briefing the unit commander and team leaders.

(6) Debriefing the unit commander and team leaders.

(7) Training of the teams.

b. The sniper team leader is responsible for the day to day activities of the sniper team. His responsibilities include:

   (1) Assuming the responsibilities of the SEO that pertain to the team in the SEO's absence.

   (2) Training the team.

   (3) Issuing necessary orders to the team.

   (4) Preparing for missions.

   (5) Controlling the team during missions.

c. Snipers work and train in two-man teams. One man's primary duty is that of the sniper, while the other serves as the observer. The sniper's weapon is the sniper weapon system. The observer has the standard service rifle, which gives the team greater suppressive fire and protection. When mounted with a night observation device, the night capability of the team is enhanced.

1-4. PERSONNEL SELECTION

Candidates for sniper training require careful screening. Commanders must screen the individual's records to determine his potential aptitude as a sniper. The rigorous training program and the increased personal risk in combat require high motivation and the ability to learn a variety of skills. Aspiring snipers must have an excellent personal record.

   a. The following are the basic guidelines to use when screening sniper candidates:

   (1) <u>Marksmanship</u>. The sniper trainee must be an expert marksman. Repeated annual qualification as expert is necessary. Successful participation in the annual competition-in-arms program and an extensive hunting background also indicate good sniper potential.

   (2) <u>Physical condition</u>. The sniper, often employed in extended operations with very little sleep, food, or water, must be in outstanding physical condition. Good health means better reflexes, better muscular control, and greater stamina. The self-confidence and control that come from athletics, especially team sports, are definite assets to a sniper trainee.

(3) **Vision**. Eyesight is the sniper's prime tool. Therefore, a sniper must have 20/20 vision or vision that is correctable to 20/20. However, wearing glasses could become a liability if they are lost or damaged. Color blindness is also considered a liability to the sniper, due to his inability to detect concealed targets that blend in with the natural surroundings.

(4) **Smoking**. A sniper should be a nonsmoker. Smoke or an unsuppressed smoker's cough can betray the sniper's position, and even though he may not smoke on a mission, refrainment may cause nervousness and irritation, which lower his efficiency.

(5) **Mental condition**. When commanders screen sniper candidates, they should look for traits that would indicate the candidate has the right qualities to be a sniper. The commander must determine if the candidate will pull the trigger at the right time and place. Some traits to look for are reliability, initiative, loyalty, discipline, and emotional stability. A psychological evaluation of the candidate can aid the commander in the selection process.

(6) **Intelligence**. Trainees must be personnel of high intelligence. A sniper's duties require a wide variety of skills. He must learn --

o Ballistics.

o Ammunition types and capabilities.

o Adjustment of optical devices.

o Radio operation and procedures.

o Observation and adjustment of mortar and artillery fire.

o Land navigation skills.

o Military intelligence collecting and reporting.

o Identification of Threat uniforms/equipment.

b. In sniper team operations involving prolonged independent employment, the sniper must also display effective decisiveness, self-reliance, good judgment, and common sense. This requires two other important qualifications; they are--

(1) _Emotional balance_. The sniper must be capable of calmly and deliberately killing targets that may not pose an immediate threat to him. It is much easier to kill in self-defense or in the defense of others than it is to kill without apparent provocation. The sniper must not be susceptible to emotions such as anxiety or remorse. Candidates whose motivation toward sniper training rests mainly in the desire for prestige may not be capable of the cold rationality that the sniper's job requires.

(2) _Fieldcraft_. The sniper must be familiar with and comfortable in a field environment. An extensive background in the outdoors and knowledge of natural occurrences in the outdoors will assist the sniper in many of his tasks. Individuals with such a background will often have great potential as a sniper.

TC 23-14

CHAPTER 2
EQUIPMENT

This chapter describes the equipment necessary for the sniper to effectively perform his mission. He carries only what is essential to successfully complete his mission. Sniper equipment may be classified as individual, team, and special.

Section I. SNIPER RIFLES

A sniper's mission requires a durable rifle with the capability of long-range precision fire. The current US Army sniper weapon system is the M21. It is being replaced by the M24 sniper weapon system.

2-1. M21 SNIPER WEAPON SYSTEM

The National Match M14 rifle (Figure 2-1) and its scope make up the M21 sniper weapon system. The rifle is accurized IAW United States Army Marksmanship Training Unit specifications and has the same basic design and operation as the standard M14 rifle (FM 23-8), except for specially selected and hand-fitted parts.

a. Differences. Significant differences are as follows:

(1) The barrels are gauged and selected to ensure correct specification tolerances. Bores are not chromium plated.

(2) The stock is walnut and impregnated with an epoxy.

(3) The receiver is individually custom fitted to the stock with a fiberglass compound.

(4) The firing mechanism is reworked and polished to provide for a crisp hammer release. Trigger weight is between 4.5 to 4.75 pounds.

(5) The suppressor is fitted and reamed to improve accuracy and eliminate any misalignment.

(6) The gas cylinder and piston are reworked and polished to improve operation and reduce carbon buildup.

(7) The gas cylinder and lower band are permanently attached to each other.

(8) Other parts are carefully selected, fitted, and assembled.

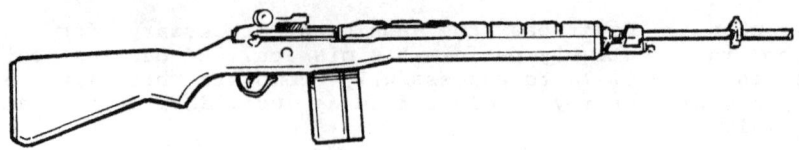

Figure 2-1. National Match M14 rifle.

b. Inspection. If the sniper discovers a deficiency while inspecting the rifle, he will report it to the unit armorer. The following areas should be inspected:

(1) Check the appearance and completeness of all parts. Shiny surfaces should be treated.

(2) Check the flash suppressor for misalignment, burrs, or evidence of bullet tipping. The suppressor should be tight on the barrel.

(3) Check the front sight to ensure that it is tight, that the blade is square, and that all edges and corners are sharp.

(4) Check the gas cylinder to ensure it fits tightly on the barrel. The gas plug should be firmly tightened.

(5) Check the forward band on the stock to ensure it does not bind against the gas cylinder front band.

(6) Check the handguard. It should not bind against the receiver, the top of the stock, or the operating rod.

(7) Check the firing mechanism to ensure the weapon will not fire with the safety "on," and that it has a smooth, crisp trigger pull when the safety is "off."

(8) Check the rear sight tension by turning the aperture up to the "10" position and then pressing down on top of the aperture with a thumb. If the aperture can be pushed down, the tension must be readjusted.

(9) Check the stock for splits or cracks.

c. **Care and Maintenance.** Extreme care has been used in building this sniper rifle. A similar degree of attention must be devoted to its daily care and maintenance.

(1) The rifle should not be disassembled by the sniper for normal cleaning and lubrication. Disassembly should be performed only by the armorer during his scheduled inspections or repair, and it will be thoroughly cleaned and lubricated at that time.

(2) The following materials are required for cleaning the rifle:

o Cleaning rod (7.62-mm, 3-piece brass, or 1-piece coated type).

o Lubricating oil.

o Bore cleaner.

o Weapon grease.

o Patches.

o Bore brush.

o Shaving brush.

o Toothbrush.

o Cleaning rags.

(3) The recommended procedure for cleaning and lubricating the rifle is as follows:

o Wipe off old oil, grease, and external dirt from the weapon.

o Clean the bore out by placing the weapon upside down on a table or in a weapon cradle. Then push a bore brush dipped in bore cleaner completely through the bore and pull it back out. Repeat this four or five times.

o Clean the chamber and bolt face with bore cleaner and a chamber brush or toothbrush.

o Clean the chamber, receiver, other interior areas, and the flash suppressor with a rag or patches.

2-3

o Wipe the bore out by pushing clean patches through the bore until they come out of the bore clean.

o Wipe off the chamber and interior surfaces with patches until clean.

o With the bolt and gas piston to the rear, place one drop of bore cleaner in between the rear band of the gas system and the lower side of the barrel. Do not put bore cleaner in the gas port!

o Lubricate the rifle by placing a light coat of grease on the operating rod handle track, camming surfaces in the hump of the operating rod, the bolt's locking lug track, and in between the front band lip of the gas system and the metal band on the lower front of the stock.

o Place a light coat of oil on all exterior metal parts.

d. Rear Sights. The M21 is equipped with National Match rear sights (Figure 2-2). The pinion assembly adjusts the elevation of the aperture. By turning it clockwise, it will raise the point of impact. Turning it counterclockwise will lower the point of impact. Each click of the pinion is 1 MOA (minute of angle) (see Appendix D). The hooded aperture is also adjustable and provides .5 MOA changes in elevation. Rotating the aperture so that the indication notch is at the top will raise the point of impact .5 MOA. Rotating the indication notch to the bottom will lower the strike of the round. The windage knob adjusts the lateral movement of the rear sight. Turning the knob clockwise will move the point of impact to the right and turning it counterclockwise will move the point of impact to the left. Each click of windage is .5 MOA.

Figure 2-2. National Match rear sight.

2-2. M24 SNIPER WEAPON SYSTEM

The components of the M24 system (Figure 2-3) are:

- o System case.
- o Bolt action rifle.
- o M3A, fixed 10x scope.
- o Scope case.
- o Detachable iron sights (front and rear).
- o Deployment case.
- o Optional bipod.
- o Cleaning kit.
- o Soft rifle case.
- o Operator's manual.

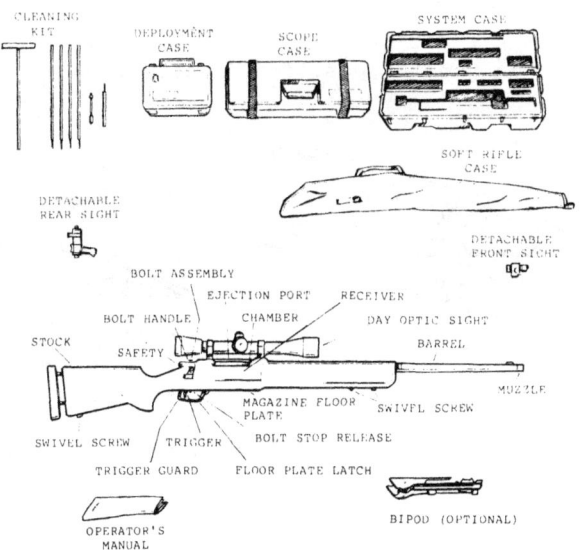

Figure 2-3. M24 sniper weapon system.

a. Rifle. The M24 is a 7.62-mm, bolt action, 5-shot repeating rifle. The rifle was designed primarily for prone shooting, but can be fired from other positions. Components of this rifle are:

o   Kevlar stock with adjustable shoulder stock.

o   Rock 5R barrel.

o   Model 40x long action with special trigger guard and floor plate assembly that allows conversion to a magnum caliber.

o   Modified model 700 trigger.

(1) The safety. The safety is located on the right rear side of the receiver and provides protection against accidental discharge under normal usage when properly engaged.

   (a) To engage the safety, place it in the "S" position (Figure 2-4).

   (b) Always place the safety in the "S" position before handling, loading, or unloading the weapon.

   (c) When the weapon is ready to be fired, place the safety in the "F" position (Figure 2-4).

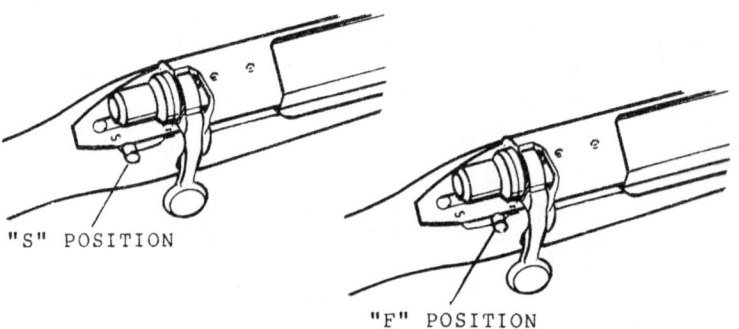

"S" POSITION

"F" POSITION

Figure 2-4. Safety.

(2) <u>Bolt assembly</u>. The bolt assembly locks the round into the chamber and extracts the round from the chamber.

(a) To remove the bolt from the receiver, place the safety in the "S" position, raise the bolt handle and pull it back until it stops. Then, push the bolt stop release up (Figure 2-5) and pull the bolt from the receiver.

BOLT STOP RELEASE

Figure 2-5. Bolt stop release.

(b) To replace the bolt, place the safety in the "S" position, align the lugs on the bolt assembly with the receiver (Figure 2-6), slide the bolt all the way into the receiver, and then push the bolt handle down.

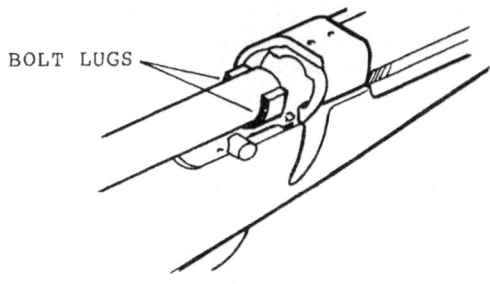

BOLT LUGS

Figure 2-6. Bolt alignment.

(3) _Trigger assembly_. Pulling the trigger fires the rifle when the safety is in the "F" position. The operator may adjust the trigger pull force from a minimum of 2 pounds to a maximum of 8 pounds. This is done using the 1/16-inch allen wrench provided in the deployment kit. Turning the trigger adjustment screw (Figure 2-7) clockwise will increase the force needed to pull the trigger. Turning it counterclockwise will decrease the force needed. This is the only trigger adjustment the sniper should make.

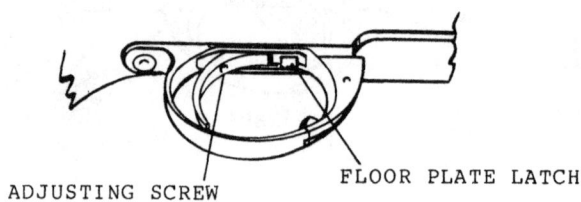

Figure 2-7. Trigger adjustment.

b. Inspection. The M24 weapon is designed to be repaired by its user. Deficiencies that cannot be repaired by the sniper will require manufacturer warranty work. Refer to TM 9-1005-306-10 that is furnished with each weapon system. The sniper must be completely familiar with this TM. The following areas should be checked when inspecting the M24:

(1) Check the appearance and completeness of all parts.

(2) Check the bolt to ensure it locks, unlocks, and moves smoothly.

(3) Check the safety to ensure it can be positively placed into "S" or "F" position easily without being too hard or moving too freely.

(4) Check the trigger to ensure the weapon will not fire when the safety is in the "S" position, and that it

has a smooth, crisp trigger pull when the safety is in the "F" position.

(5) Check the guard screws for proper torque (65 inch-pounds).

(6) Check the scope mounting ring nuts for proper torque (65 inch-pounds).

(7) Check the stock for any cracks, splits, or any contact it may have with the barrel.

(8) Inspect the scope for obstructions, such as dirt, dust, moisture, or loose or damaged lenses.

c. Care and Maintenance. The M24 does not require the same amount of maintenance as the M21; but it does require some.

(1) The following materials are required for cleaning the rifle:

o Cleaning rod (7.62-mm, 3-piece brass, or 1-piece coated type).

o Lubricant (CLP/LSA).

o Rifle bore cleaner (RBC).

o Patches.

o Bore and chamber brushes.

o Toothbrush.

o Cleaning rags.

(2) The recommended procedure for cleaning and lubricating the rifle is as follows:

o Remove the bolt assembly from the receiver and push the floor plate latch (Figure 2-7) to release the floor plate.

o Lay the weapon on a table or in a weapon cradle with the barrel laying lower than the receiver, and the ejection port facing down.

o Attach a bore brush to the cleaning rod and dip it in bore cleaner.

o Push the bore brush all the way through the bore from the chamber end of the rifle and then pull it back through the bore. Repeat this four or five times.

o Clean the chamber with a chamber brush and bore cleaner until it is clean.

o Attach a cleaning tip with a patch and push it through the bore from the chamber end of the rifle.

o Repeat this with clean patches until the patches come out clean.

o Clean the bolt face with a brush and bore cleaner.

o Wipe clean the interior of the receiver and magazine with cloth or patches.

o Pull a piece of paper or thin plastic underneath the bottom of the barrel in the groove between the barrel and the forestock. Keeping it inside this space, pull it all the way to the chamber and then push it back out. This cleans out any obstruction and ensures no contact is made between the stock and barrel.

o Put a thin layer of lubricant on the bolt lugs and cocking cams. The exterior metal surfaces of the weapon have been specially treated and require no coating with any lubricant unless the weapon is to be stored for a long time.

o Apply a thin coat of lubricant on the bore, chamber, bolt face, and the exterior of the trigger guard assembly if the weapon is to be stored. Before firing, these areas must all be wiped dry.

d. Disassembly. Occasionally the weapon will require disassembly; however, this should be done only when absolutely necessary, not for daily cleaning. An example of this would be to remove an obstruction that is stuck between the forestock and the barrel. When disassembly is required, the recommended procedure is as follows:

o Place the weapon so it is pointing in a safe direction.

o Ensure the safety is in the "S" position.

o Remove the bolt assembly.

o Loosen the mounting ring nuts (2) (Figure 2-8) on the scope and remove the scope.

o Remove the trigger guard screws (2) (Figure 2-9).

o Lift the stock from the barrel assembly (Figure 2-10).

o For further disassembly, refer to TM 9-1005-306-10.

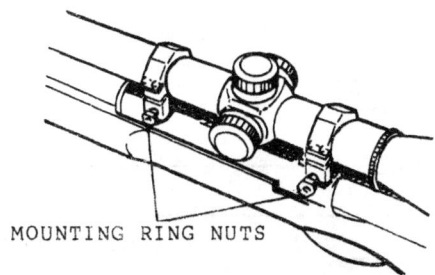

Figure 2-8. Ring nuts.

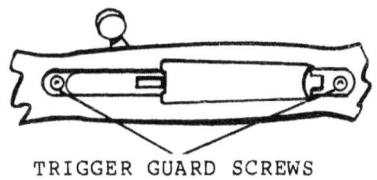

Figure 2-9. Trigger guard screws.

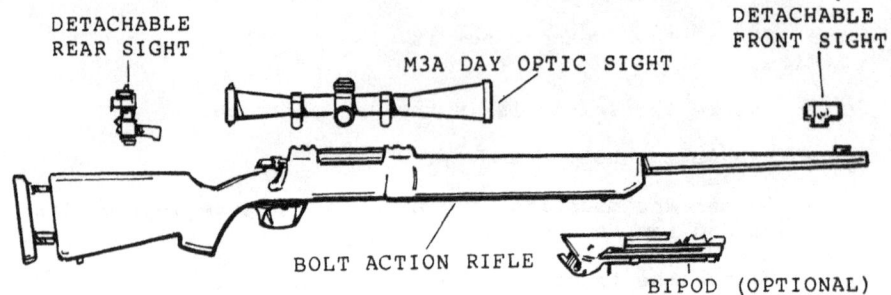

(a) MAJOR ASSEMBLIES

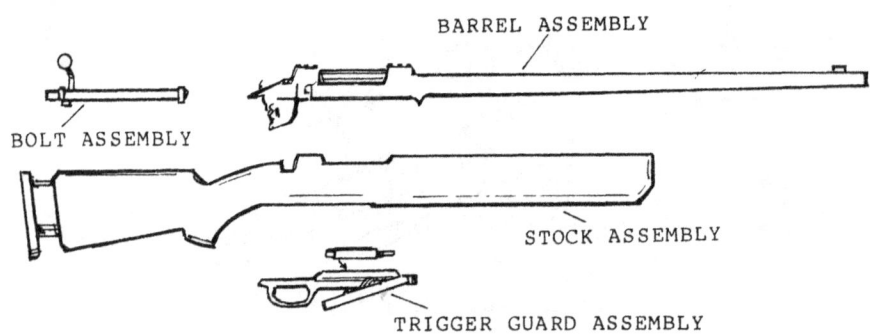

(b) RIFLE

Figure 2-10. Disassembled weapon.

e. Iron Sights. The M24 has detachable front and rear iron sights, which give the sniper a back-up sighting system.

(1) To attach the front sight to the barrel, align the front sight and the front sight base dovetails and slide the sight over the base. Next, tighten the screw slowly, ensuring the screw seats into the recess in the sight base (Figure 2-11).

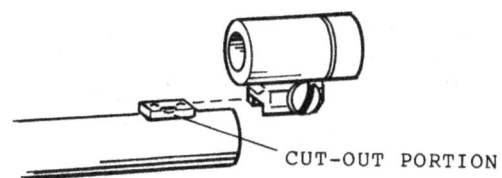

Figure 2-11. Front sight attachment.

(2) To attach the rear sight to the receiver, remove one of the three set screws, align the rear sight with the rear sight base located on the left rear of the receiver (Figure 2-12). Tighten the screw to secure the sight to the base.

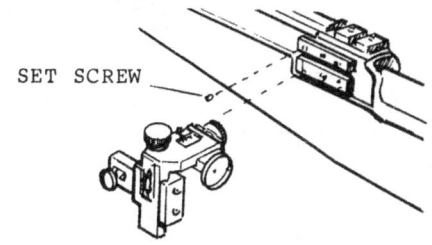

Figure 2-12. Rear sight attachment.

(3) Adjustments of elevation are made by turning the elevation knob located on the top of the rear sights. Turning the knob in the direction marked "UP" will raise the point of impact. Turning in the opposite direction will lower the point of impact. Each click of adjustment equals .25 MOA.

(4) Adjustments in windage are made by turning the windage knob located on the right side of the rear sights. Turning the knob in the direction marked "R" will move the point of impact to the right. Turning it in the opposite direction will move the point of impact to the left. Each click of adjustment equals .25 MOA.

(5) There are 12 divisions, or 3 MOA adjustments in each knob revolution. Total elevation adjustment latitude is 60 MOA and 36 MOA windage adjustments. Adjustment scales are of the "vernier" type. Each graduation on the scale plate equals 3 MOA. Each graduation on the sight base scale equals 1 MOA. To use, note the point at which graduations on both scales are aligned. Count the number of full 3 MOA graduations from "0" on the

scale plate to "0" on the sight base scale. Add this figure to the number of 1 MOA from "0" on the bottom scale to the point where the two graduations are aligned (see Figure 2-13). After zeroing the sight to the rifle

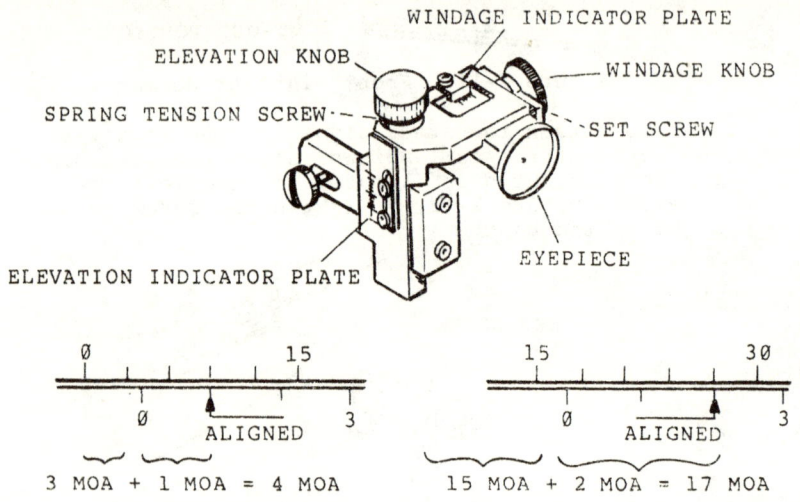

Figure 2-13. Rear sight adjustment.

at the preferred range, loosen the elevation and windage indicator plate screws with the hex wrench provided. Align the "0" on the plate with the "0" on the sight body. Retighten the plate screws. Now, loosen the set screws in each knob and align the "0" of the knob with the reference line on the sight. Pressing the knob against the sight, tighten the set screws. The click can be sharpened or softened to your preference by loosening or tightening the spring screws equally on each knob. You can now make windage and elevation corrections and return quickly to "zero" standard.

f. Loading. The M24 has an internal, 5-round capacity magazine. To load the rifle --

o Point the rifle in a safe direction.

o Ensure the safety is in the "S" position.

o Raise the bolt handle and pull it back until it stops.

o Push five rounds of 7.62-mm special ball ammunition one at a time through the ejection port into the magazine. The bullet end of the rounds should be aligned toward the chamber.

o Push the rounds fully rearward in the magazine.

o Once the five rounds are in the magazine, push the rounds downward while slowly pushing the bolt forward over the top of the first round.

o Push the bolt handle down. The magazine is now loaded.

o To chamber a round, raise the bolt and pull it back until it stops.

o Push the bolt forward. The bolt will remove a round from the magazine and push it into the chamber.

o Push the bolt handle down.

o To fire, place the safety in the "F" position and pull the trigger.

Section II. SNIPERSCOPES

A sniperscope mounted on the rifle allows the sniper to detect and engage targets more effectively than he could by using the iron sights. Unlike sighting with iron sights, the target's image in the scope is in focus with the aiming point (reticle). This allows for a more focused picture of the target and aiming point at the same time. Another advantage of the scope is its ability to magnify the target. This increases the resolution of the target's image, making it clearer and more defined. Keep in mind, a scope does not make you shoot better, it only helps you see better.

2-3. AUTO-RANGING TELESCOPE

Auto-ranging telescopes are part of the M21 system. There are two types of ARTs found on the M21 system; the ART I and the ART II. The basic design and operating principle of both scopes are the same. Therefore, they will be described together, but their differences will be pointed out.

a. Components. The ART has a commercially procured 3 to 9 variable power telescopic sight, modified for use with the sniper rifle. This scope has a modified reticle with a ballistic cam mounted to the power adjustment ring on the ART I (Figure 2-14). The ART II (Figure 2-15) has a separate ballistic cam and power ring. The ART is mounted on a spring-loaded base mount that is adapted to fit the M14. It comes with a hard carrying case used to transport it when it is not mounted to the rifle.

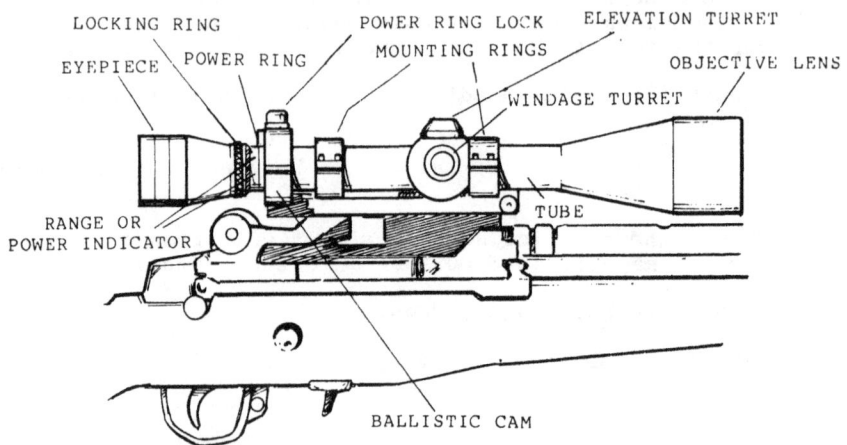

Figure 2-14. ART I scope.

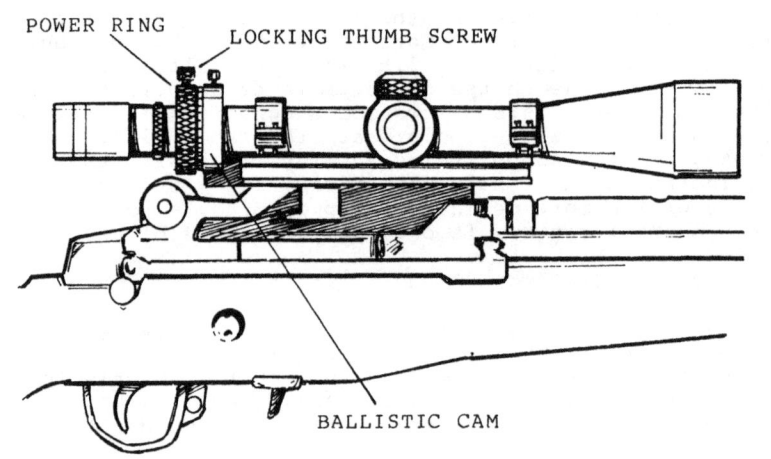

Figure 2-15. ART II scope.

b. Magnification. The ART's increased magnification allows the sniper to see the target clearer.

(1) The average unaided human eye can distinguish detail of about 1 inch at 100 yards (1 minute of angle). Magnification, combined with well-designed optics, permit resolution of this 1 inch divided by the magnification. Thus a 1/4 MOA of detail can be seen with a 4x scope at 100 yards or 1 inch of detail can be seen at 600 yards with a 6x scope.

(2) The lens surfaces are coated with a hard film of magnesium fluoride for maximum light transmission.

(3) Located midway on the scope tube are the elevation and windage turrets with dials that are used for zeroing adjustments. These dials are graduated in .5 MOA increments.

(4) These telescopes also have modified reticles. The ART I scope has the basic cross hair design reticle with two vertical stadia lines that appear at target distances, 15 inches above and 15 inches below the horizontal line of the reticle (Figure 2-16). It also has two horizontal stadia lines that appear at target distance, 30 inches to the left and 30 inches to the right of the vertical line of the reticle.

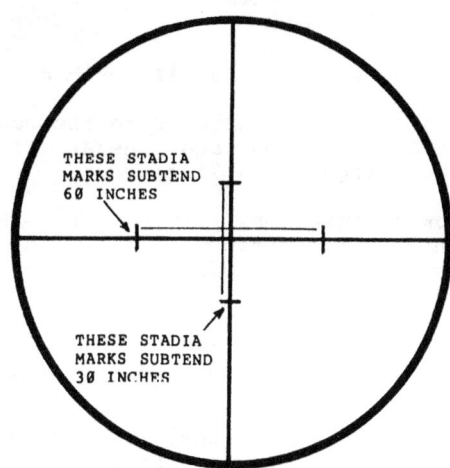

Figure 2-16. ART I reticle.

(5) The ART II scope reticle (Figure 2-17) consists of three posts; two horizontal and one bottom vertical post. These posts represent 1 meter at the target's distance. The reticle has a basic cross hair with two dots on the horizontal line that appear at target distance, 30 inches to the left and 30 inches to the right of the vertical line.

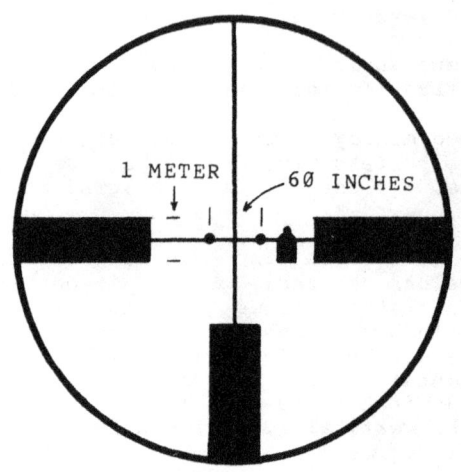

Figure 2-17. ART II reticle.

(6) A ballistic cam is attached to the power adjustment ring on the ART I scope, and the ART II scope has a separate power ring and ballistic cam.

(7) The power ring on both scopes increases and decreases the magnification of the scope, while the ballistic cam raises and lowers the scope to compensate for elevation.

(8) Focus adjustments are made by screwing the eyepiece into or away from the scope tube until the reticle is clear.

c. Scope Mount. The ART mounts are made of lightweight aluminum consisting of a side-mounting plate and a spring-loaded base with scope mounting rings. The mount is designed for low profile mounting of the scope to the rifle, using the mounting guide grooves and threaded hole(s) on the left side of the receiver. The ART I has one thumb screw that screws into the left side of the receiver (Figure 2-18). The ART II mount has two thumb screws; one is screwed into the left side of the receiver, and the other is screwed into the cartridge clip guide in front of the rear sight (Figure 2-19).

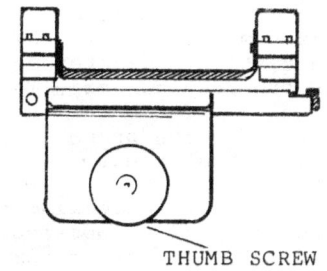

Figure 2-18. ART I mount.

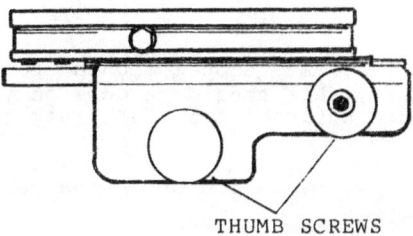

Figure 2-19. ART II mount.

d. Design and Operation. The ART scopes are designed to automatically adjust for the needed elevation at ranges of 300 to 900 meters. This is done by increasing or decreasing the magnification of the scope until a portion of the target's image matches the represented measurement of the scope's reticle.

(1) For example, adjust the power ring on the ART I scope until 30 inches of an object or a person's image (beltline to top of head) fits exactly in between the

vertical stadia lines (top stadia line touching top of the head and bottom stadia line on the beltline).

(2) Another example of this is to adjust the power ring on the ART II scope until 1 meter (approximately 40 inches) of a person or an object's image appears equal to one of the posts in the reticle.

(3) When turning the power ring to adjust the target's image to the reticle, the ballistic cam is also being turned. This raises or lowers the scope itself to compensate for elevation. Therefore, once the scope's magnification is properly adjusted in proportion to the target's image, the ballistic cam has at the same time adjusted the scope for the proper elevation needed to engage the target at that range.

(4) The ART II scope has a locking thumbscrew located on the power ring used for connecting and disconnecting the power ring from the ballistic cam. This allows the sniper to adjust the scope on the target (auto-ranging mode) and then disengage the locking thumbscrew to increase magnification (manual mode) without affecting the elevation adjustment.

e. Zeroing. The ART scope should be zeroed at 300 meters. Ideally, this should be done on a known-distance range, with international bull-type targets. When zeroing the ART scope --

(1) Remove the elevation and windage adjustment caps from the scope.

(2) Turn the power adjustment ring to the lowest position (3). On the ART II scope, ensure the locking thumbscrew is engaged and that the ballistic cam moves when the power ring is turned.

(3) Assume a good prone supported position that allows the natural point of aim to be centered on the target.

(4) Fire three rounds, using good marksmanship fundamentals with each shot.

(5) Make the needed adjustments to the scope after placement of the rounds has been noted (Figure 2-20). Be sure you remember --

o That each mark on the elevation and windage dials equals .5 MOA. (.5 MOA at 300 meters equals 1.5 inches.)

o That turning the elevation dial in the direction of the UP arrow will raise the point of impact; turning it the other direction will lower it.

o That turning the windage dial in the direction of the R arrow will move the point of impact to the right; turning it the other direction will move it to the left.

WINDAGE SCALE - INTERNAL ADJUSTMENT
RIGHT SIDE

ELEVATION SCALE - INTERNAL ADJUSTMENT
TOP

Figure 2-20. Elevation and windage scales.

(6) Repeat the steps in (4) and (5) above until two 3-round shot groups are centered on the target.

(7) After the scope is properly zeroed, it will effectively range on targets out to 900 meters in the auto-ranging mode.

2-4. LEUPOLD M3A TELESCOPE

The M3A telescope is mounted on the M24 sniper weapon system. The design and operating principle of the M3A scope are different than the ART series of scopes. The most noticeable difference in the M3A is the method that is used to adjust the scope for varying distances.

    a. Components. The M3A consists of the telescope, a fixed mount, extendable sun shade for the objective lens, and dust covers for the objective and eyepiece lens.

    b. Magnification. The telescope has a fixed 10-power magnification, which gives the sniper better resolution than found with the ART series of scopes.

        (1) There are three knobs located midway on the tube. These are the focus, elevation, and windage knobs (Figure 2-21).

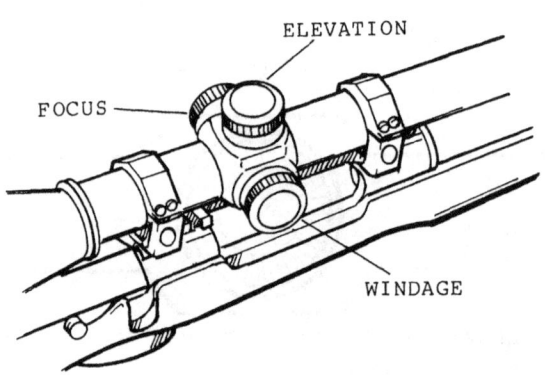

Figure 2-21. Adjustment knobs.

    o The focus knob is used to focus the target's image on the same focal plane as the reticle, thereby reducing parallax to a minimum. The focus knob has two extreme positions indicated by the infinity mark and the largest of four dots. Adjustments between

2-22

these positions will focus images from less than 50 meters to infinity.

o The elevation knob is located on top of the tube. This knob has calibrated index markings from 1 to 10. These markings represent the elevation setting adjustments needed at varying distances; 1 = 100, 3 = 300, 7 = 700 meters, and so on. Each click of the elevation knob equals 1 MOA.

o The windage knob is located on the right side of the tube. This knob is used to make lateral adjustments to the scope. Turning the knob in the indicated direction will move the point of impact in that direction. Each click on the windage knobs equals .5 MOA.

(2) The eyepiece is adjusted by turning it in or out of the tube until the reticle appears crisp and clear (Figure 2-22). Focusing the eyepiece should be done after mounting the scope. Grasp the eyepiece and back it away from the lock ring. Do not attempt to loosen the lock ring first; it will automatically be loose when you back away the eyepiece (no tools are needed). Turn the eyepiece several turns so as to move it at least 1/8 inch. It will take this much change to achieve any measurable effect on the focus. Look through the scope at the sky or a blank wall and check to see if the reticle appears sharp and crisp.

o The reticle is a duplex style, mil dot reticle that has thick outer sections and a thin center section (Figure 2-23). Superimposed on the thin center section is a series of dots; four on each side of center and four above and below center. Each of these dots are spaced one mil apart and one mil away from both the center and the thick outer sections.

o These mil dots are used to estimate distances to targets. To do this, the sniper must first know the size of the target at the given distance. Once this is known, the sniper simply compares the size of the target's image with the spacing between the mil dots on the reticle. The sniper then uses the mil relation formula (see Chapter 4) to determine the distance to the target.

2-23

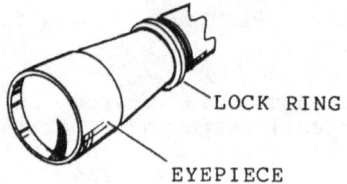

Figure 2-22. Eyepiece adjustment.

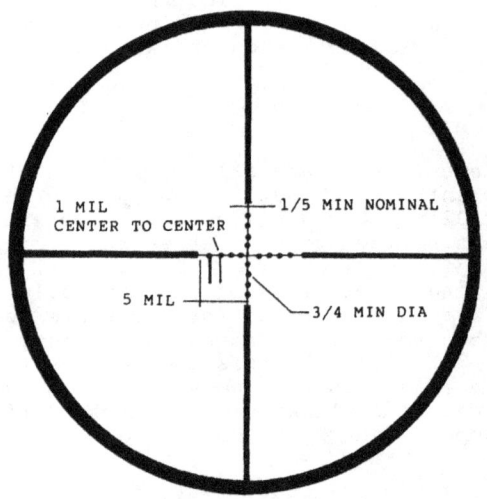

Figure 2-23. Reticle.

c. Scope Mount. The scope mount has a base plate with four screws; a pair of scope rings with eight ring screws, each with an upper and lower ring half; and two ring mounting bolts with nuts (Figure 2-24). The base plate is mounted to the rifle by screwing the four base plate screws through the plate and into the top of the receiver. Ensure that the screws do not protrude into the receiver and interrupt the functioning of the bolt. After the base plate is mounted, the scope rings are mounted. To mount the scope rings, select one set of slots on the mounting base and engage each ring bolt spline with the selected slot. Next, slide the mount claw against the base and finger tighten the mount ring nut. Then check the eye relief. If the scope needs to be adjusted, loosen the ring nuts and align the ring bolts with the other set of slots on the base and repeat the process again. Once the sniper is satisfied with the eye relief obtained, he will then tighten the ring nuts to 65 inch-pounds using the T-handle torque wrench.

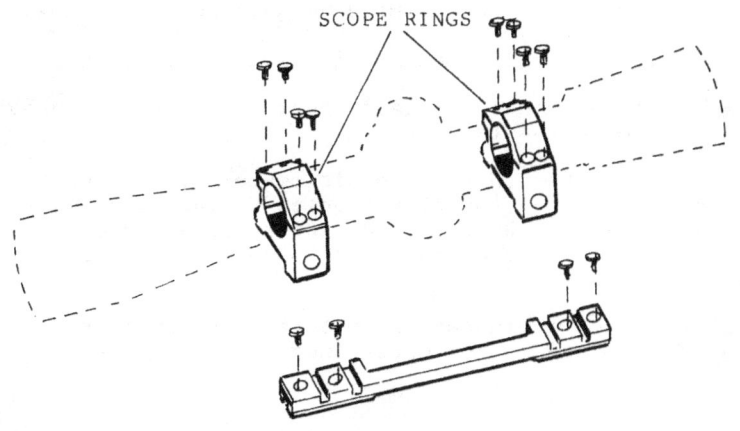

Figure 2-24. Scope mount.

d. Operation. When using the scope, the sniper simply looks at the target, determines the distance to it by using the mil dots on the reticle, and then adjusts the elevation knob for the given range.

e. Zeroing. Zeroing the M3A scope should be done on a known-distance range (preferably 900 meters long) with international bull-type targets. When zeroing the scope --

(1) Assume a good prone supported position 100 meters away from the target.

(2) Ensure the elevation knob is lined up on the index line marked "1."

(3) Fire three rounds at the center of the target, keeping the same aiming point each time.

(4) After placement of the rounds has been noted, turn the elevation and windage knobs to make the needed adjustments to the scope:

- o Each click on the elevation knob equals one MOA.

- o One MOA at 100 meters equals about 1 inch.

- o Each click on the windage knob equals .5 MOA.

- o .5 MOA at 100 meters equals about .5 inch.

(5) Repeat this process until a 3-round group is centered on the target.

(6) Once the shot group is centered, loosen the hex head screws on the elevation and windage dials. Turn the elevation knob to the index line marked "1" (if needed). Turn the windage knob to the index line marked "0" (if needed).

(7) After zeroing at 100 meters, confirm this zero out to 900 meters at 100-meter increments.

Section III. AMMUNITION

The sniper uses the 7.62-mm special ball (M118) with the sniper weapon systems. The sniper must rezero his weapon each time he fires a different lot number of ammunition. This information should be maintained in the weapon's data book. (See Appendix C.)

2-5. SPECIAL BALL

The 7.62-mm, M118 special ball cartridge consists of a gilding metal jacket and a lead antimony slug. It is a boat-tailed bullet (rear of bullet is tapered) and weighs 173 grains. The tip of the bullet is not colored. The base of the cartridge is stamped with a circle that has a vertical and horizontal line sectioning it in quarters along with the year of manufacture. Its primary use is against personnel. It has an extreme spread (accuracy standard) for a ten-shot group of no more than 12 inches at 550 meters (fired from an accuracy barrel in a test cradle).

2-6. BLANKS

M82, 7.62-mm blank ammunition is used during sniper field training. This ammunition provides the muzzle blast and flash that can be detected by trainers during the exercises that evaluate the sniper's ability to conceal himself while firing his weapon.

Section IV. OPTICAL OBSERVATION EQUIPMENT

The sniper's success in selecting and engaging targets without betraying himself depends upon his powers of observation. In addition to the sniperscope, the sniper team has an observation telescope, binoculars, night vision sight, and night vision goggles to enhance their ability to observe and engage targets. Team members must relieve each other when using this equipment since prolonged use will cause eye fatigue, significantly reducing the effectiveness of observation. Periods of observation during daylight should be limited to 30 minutes followed by at least 15 minutes of rest. When using night vision devices, limit the observer's initial period of viewing to 10 minutes followed by a 15-minute rest period. After several periods of viewing, extend the viewing period to 15 and then 20 minutes.

2-7. OBSERVATION TELESCOPE

The M49 observation telescope is a prismatic optical instrument of 20-power magnification (Figure 2-25). Components of the telescope include a removable eyepiece and objective lens covers, an M15 tripod with canvas carrier, and a hard case carrier for the telescope. The telescope is focused by turning the eyepiece in or out until the image of the object being viewed is crisp and clear to the viewer. The sniper team carries the telescope on all missions. The observer uses the telescope to determine wind speed and direction by reading mirage (Chapter 3), observing the bullet trace, and observing the bullet impact. This information is used by the sniper to make quick and accurate adjustments for wind conditions. The lenses are coated with a hard film of magnesium fluoride for maximum light transmission. Its high magnification makes observation, target detection, and target identification possible where conditions and range would otherwise preclude this capability. Camouflaged targets and those in deep shadows can be more readily distinguished. The team can observe troop movements at greater distances and identify selective targets with ease.

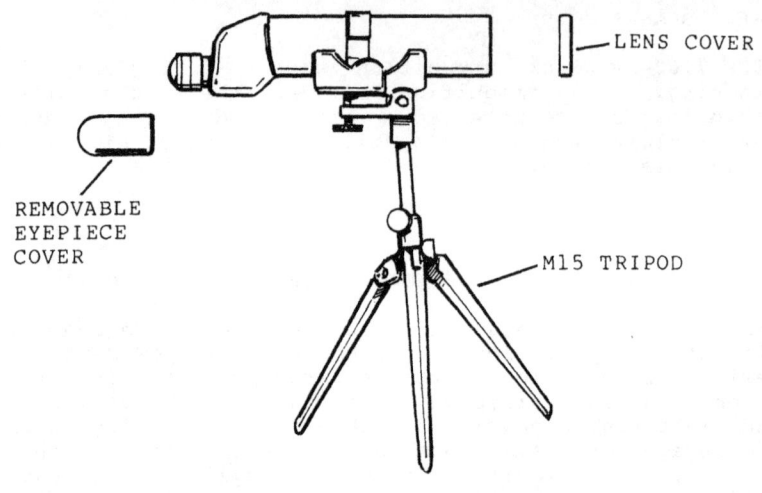

Figure 2-25. M49 observation telescope.

2-8. BINOCULARS

The M19 binoculars (Figure 2-26) have 7-power magnification with 50-mm objective lenses. The M19 has an interpupillary scale located on the hinge. The sniper should adjust the binoculars until one sharp circle appears while looking through them. After adjusting the binoculars' interpupillary distance (distance between a person's pupils), the sniper should make a mental note of the reading on this scale for future reference. The eyepieces are also adjustable. The sniper will adjust one eyepiece at a time by turning the eyepiece with one hand while placing the palm of the other hand over the objective lens of the other monocular. While keeping both eyes open, he will adjust the eyepiece until he can see a crisp, clear view. After one eyepiece is adjusted, he will repeat the procedure again with the remaining eyepiece. The sniper should also make a mental note of the diopter scale reading on both eyepieces for future reference. One side of the binoculars has a laminated reticle pattern (Figure 2-27) that consists of a vertical and horizontal mil scale that is graduated into 10-mil increments. Using this reticle pattern aids the sniper in determining range and adjusting indirect fires. The sniper uses the binoculars for --

o   Calling for and adjusting indirect fires.

o   Observing target areas.

o   Observing enemy movement and positions.

o   Identifying aircraft.

o   Improving low-light level viewing.

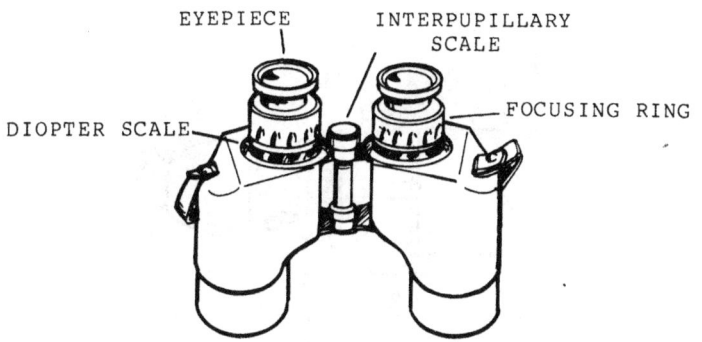

Figure 2-26. M19 binoculars.

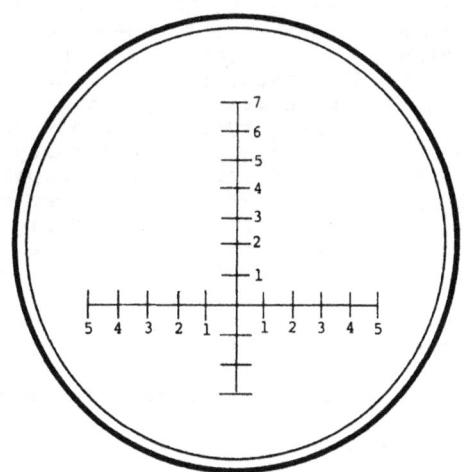

Figure 2-27. M19 reticle.

2-9. NIGHT VISION SIGHT

The night vision sight (AN/PVS-4) is a portable, battery-operated, electro-optical instrument that can be hand-held for visual observation or weapon-mounted for precision fire at night (Figure 2-28). The observer can detect and resolve distant targets through the unique capability of the sight to amplify reflected ambient light (moon, stars, or sky-glow). The sight is passive; thus, it is free from enemy detection by visual or electronic means. This sight, with appropriate weapons adapter bracket, may be mounted on the M16 rifle.

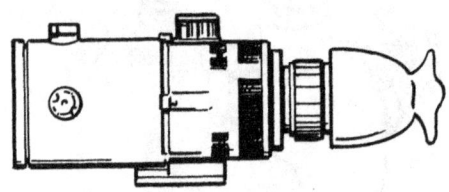

Figure 2-28. AN/PVS-4 night vision sight.

a. Uses. When mounted on the M16 rifle, the AN/PVS-4 is very effective in achieving a first-round hit out to 300 meters, depending on the light conditions. The AN/PVS-4 is mounted on the M16 because the limited range of the night sight does not make its use practical for the sniper weapon system. This also avoids problems that may occur when removing and replacing the sniperscope. The night sight provides an effective observation capability during night combat operations. The sight does not give the width, depth, or clarity of daylight vision; however, a well-trained operator can see enough to analyze the tactical situation, detect enemy targets, and place effective fire on them. The sniper team uses the AN/PVS-4 to --

o Enhance their night observation capability.

o Locate and suppress hostile fire at night.

o Deny enemy movement at night.

o Demoralize the enemy with effective first-round kills at night.

b. Employment Factors. Since the sight requires target illumination and does not project its own light source, it will not function in complete darkness. The sight works best on a bright, moonlit night. When there is no light or the ambient light level is low (such as in heavy vegetation), the use of artificial or infrared light will improve its performance.

(1) Fog, smoke, dust, hail, or rain limit the range and decrease the resolution of the instrument.

(2) The sight cannot see through objects in the field of view. For example, the operator will experience the same range restrictions when viewing dense woodlines as he would when using other optical sights.

(3) Initially, an operator may experience eye fatigue when viewing for prolonged periods. Initial exposure should be limited to 10 minutes followed by a rest period of 15 minutes. After several periods of viewing, he can safely extend this time limit. To aid in maintaining a continuous viewing capability and to reduce eye fatigue, it is recommended that the operator frequently alternate his viewing eyes.

c. Zeroing. The operator may zero the sight during pure daylight or darkness; however, he may have some difficulty in zeroing just before darkness. The light level at dusk is too low to permit the operator to resolve his zero target with the lens cap cover in place, but it is still intense enough to cause the sight to automatically cut off unless the lens cap cover is in position over the objective lens. The sniper will normally zero the sight for the maximum practical range that he can be expected to observe and fire, depending on the level of light.

2-10. NIGHT VISION GOGGLES (AN/PVS-5)

These are a lightweight, passive night vision system that gives the sniper team another means of observing an area during darkness (Figure 2-29). The goggles are normally carried by the sniper because the observer has the M16 mounted with the night sight. The goggles are easier to view with because of their design. However, the same limitations that apply to the night sight also apply to the goggles.

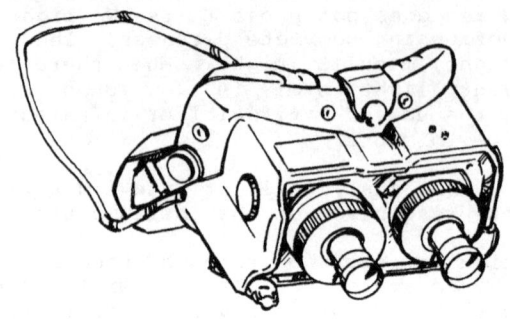

Figure 2-29. AN/PVS-5 night vision goggles.

Section V. CLOTHING AND ADDITIONAL EQUIPMENT

The sniper must use special clothing and equipment to reduce the possibility of detection.

2-11. CAMOUFLAGE

The sniper is outfitted with the following standard camouflage clothing and equipment designed for tropical or temperate zones. He may use natural or artificial materials to modify the clothing to match his environment.

    a. Ghillie Suit. The sniper constructs a ghillie suit by modifying a set of camouflage fatigues. (See Chapter 4.) He adds canvas to the front and elbows for protection when crawling. He also attaches garnish to the back and shoulders to break up his outline and to help him to blend into the surrounding terrain.

    b. Hat. The camouflaged hat is considered part of the sniper's ghillie suit and should be camouflaged with the same materials. (See Chapter 4.)

    c. Boots. The sniper should wear boots that are suitable to the climate he is in. The boots need to be camouflaged in the same manner as the ghillie suit.

d. Rucksack. At a minimum, the sniper's rucksack will contain a two-quart canteen, an entrenching tool, a first aid kit, a lensatic compass, pruning shears, a sewing kit with canvas needles and nylon thread, spare netting and garnish, rations, and personal items as needed. The sniper will also carry his ghillie suit in his rucksack until the mission requires him to wear it.

2-12. ADDITIONAL EQUIPMENT

Additional equipment the sniper may use for normal or special missions includes the following:

   a. Communications Equipment. The sniper team must have a man-portable radio that will give the team secure communications with units involved in their mission.

   (1) AN/PRC-77. The basic radio for the sniper team is the AN/PRC-77 (Figure 2-30). This radio is a short-range, man-pack portable, frequency modulated receiver-transmitter that provides two-way voice communication. The set is capable of netting with all other infantry and artillery FM radio sets on common frequencies. The AN/KY-57 should be installed with the AN/PRC-77. This allows the sniper team to communicate securely with all units supporting or being supported by the sniper team.

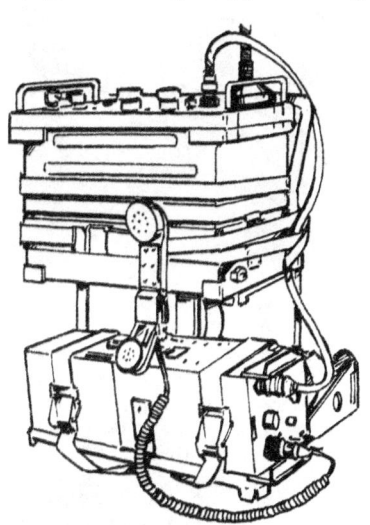

Figure 2-30. AN/PRC-77 radio.

(2) AN/PRC-119. The AN/PRC-119 is currently replacing the AN/PRC-77. This radio is a manpack portable, VHF/FM radio that is designed for simple, quick operation using a 16-element key pad for push-button tuning. It is capable of short-range and long-range operation for voice, FSK, or digital data communications. It can be used for single-channel operation or in a jam-resistant, frequency-hopping mode, which can be changed as needed. This radio has a built-in self-test with a visual and audio readback. It is compatible with the AN/KY-57 for secure communications.

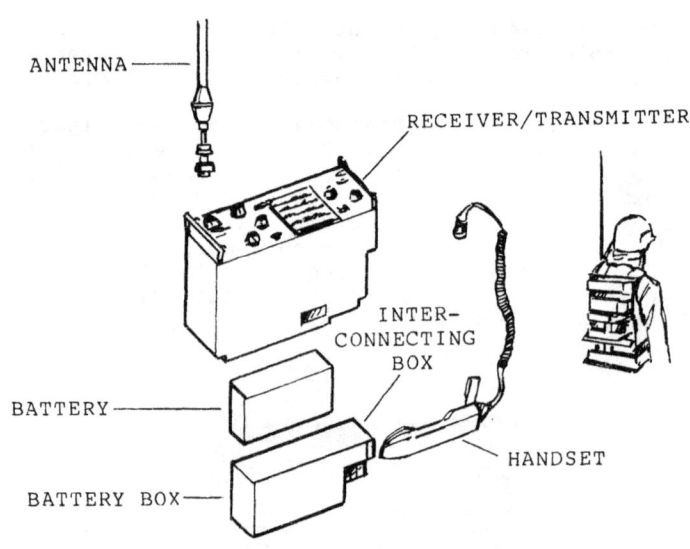

Figure 2-31. AN/PRC-119 radio.

b. Sidearms. Each member of the team should have a sidearm, such as an M9, 9-mm Beretta or a caliber .45 pistol. A sidearm gives a sniper the needed protection from a nearby threat while on the ground moving or while in the confines of a sniper position.

c. Compass. Each member of the sniper team must have a lensatic compass for land navigation. The team must have military maps of the area they are operating in.

d. **Calculator.** The sniper team will need a pocket-size calculator to calculate distances when using the mil-relation formula. Solar-powered calculators usually work fine, but under low light conditions, battery power may be preferred. If a battery-powered calculator is to be used in low light conditions, it should have a lighted display.

e. **AN/GVS-5, Laser Observation Set.** Depending on the mission, snipers may use the AN/GVS-5 to determine increased distances. The AN/GVS-5 (LR) (Figure 2-32) is an individually operated, hand-held distance measuring device designed for distances from 200 to 9,990 meters (+/- 10 meters). It measures distances by firing an infrared beam at a target and measuring the time the reflected beam takes to return to the operator. It then displays the target distance in meters inside the viewer. The reticle pattern in the viewer is graduated in 10-mil increments and has display lights to indicate low battery and multiple target hits. If the beam hits more than one target, the display will give a reading of the closest target hit. The beam that is fired from the set poses a safety hazard. Snipers planning to use this equipment should be thoroughly trained in its safe operation. (See TM 11-5860-201-10.)

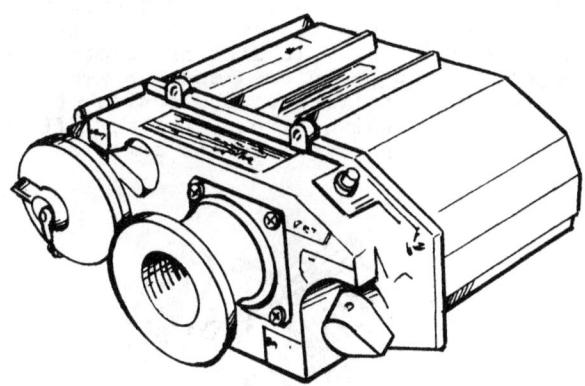

Figure 2-32. Laser observation set.

f. **AN/PVS-7A, Night Vision Goggles.** The AN/PVS-7A night vision goggles (Figure 2-33) can be used instead of the AN/PVS-5 goggles. These goggles have a better resolution and viewing capability than the AN/PVS-5 goggles. The

AN/PVS-7A goggles come with a head mount assembly that allows them to be mounted in front of the face, allowing both hands to be free. The goggles can be used without the mount assembly for hand-held viewing. (See TM 11-5855-262-10-1.)

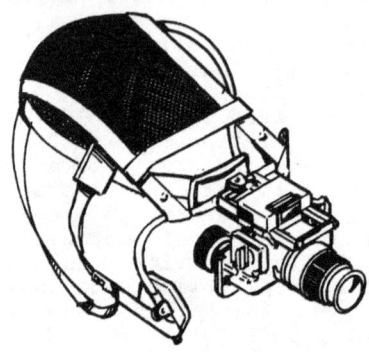

Figure 2-33. Night vision goggles.

g. M22 Binoculars. The M22 binoculars (Figure 2-34) can be used instead of the M19. These binoculars have the same features as the M19, plus fold-down eyepiece cups for personnel who wear glasses to reduce the distance between the eyes and the eyepiece. It also has protective covers for the objective and eyepiece lenses. The binoculars have laser protection filters on the inside of the objective lenses (direct sunlight can reflect off of these lenses). The reticle pattern is the same as in the M19 binoculars.

Figure 2-34. M22 binoculars.

TC 23-14

CHAPTER 3
SNIPER MARKSMANSHIP

Sniper marksmanship is an extension of basic rifle marksmanship and focuses on the techniques needed to engage targets at extended ranges. To successfully engage targets at these increased distances, the sniper must be proficient in sniper marksmanship fundamentals and a variety of other areas. Examples of these areas are determining the effects of weather conditions on ballistics; holding off for elevation and windage; engaging moving targets; using and adjusting scopes; and zeroing procedures. As with all sniper skills, sniper marksmanship is a perishable skill that must be practiced often.

3-1. USING THE FUNDAMENTALS OF MARKSMANSHIP

A sniper must be thoroughly trained in the fundamentals of marksmanship. These include assuming a position, aiming, breath control, and trigger control. These fundamentals develop fixed and correct shooting habits for instinctive application. Every sniper should periodically refamiliarize himself with these fundamentals regardless of his experience.

 a. Assuming a Firing Position. The sniper should fire from a prone supported position (Figure 3-1). Only when a prone supported position cannot be used will the sniper use an alternate type of position. In any type of position, the sniper should always use artificial support for the weapon. This can be sandbags, rucksacks, logs, or anything that will provide a stable platform for the rifle. This reduces movement of the weapon caused by contact with the body. First shot accuracy is an absolute must for the sniper's mission. There are five elements common to a good firing position.

3-1

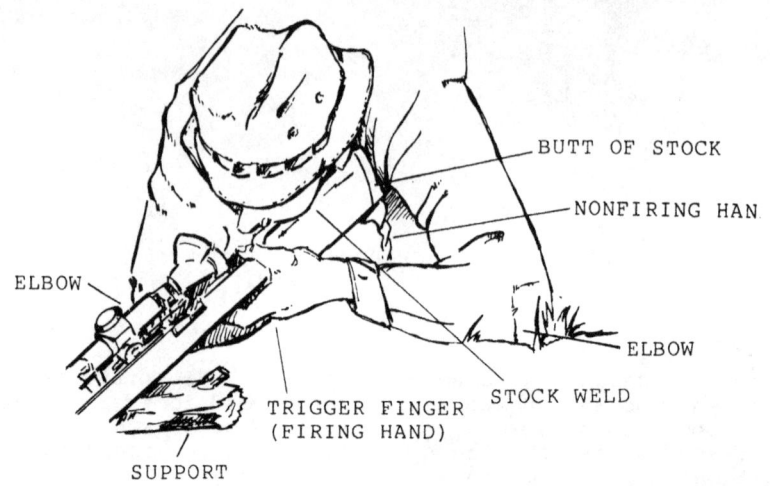

Figure 3-1. Firing position.

(1) <u>Nonfiring hand</u>. Use the nonfiring hand to support the butt of the weapon. The sniper places his hand next to his chest and rests the tip of the butt on it. He balls his hand into a fist to raise the weapon's butt or loosens the fist to lower the weapon's butt. A preferred method to do this is to hold a sock full of sand in the nonfiring hand and place the weapon butt on the sock. This reduces body contact with the weapon. To raise the butt, the sniper squeezes the sock and to lower it, he loosens his grip on the sock.

(2) <u>Butt of the stock.</u> Place the butt of the stock firmly in the pocket of the shoulder. The sniper can insert a pad on the ghillie suit where contact with the butt is made. This reduces pulse beat and breathing effects that can be transmitted to the weapon.

(3) <u>Firing hand</u>. With the firing hand, grip the small of the stock. Using the middle through little fingers, exert a slight rearward pull to keep the butt of the weapon firmly in the pocket of the shoulder. Place the thumb over the top of the small of the stock. Place the index finger on the trigger, ensuring it does not touch the stock of the weapon and will not disturb the lay of the rifle when the trigger is pulled.

(4) **Elbows**. Find a comfortable position that provides the greatest support.

(5) **Stock weld**. The sniper needs to ensure he places his cheek in the same place on the stock with each shot. A change in stock weld tends to cause misalignment of sights, thus creating misplaced shots.

b. Aiming the Rifle. Begin the aiming process by aligning the rifle with the target when assuming a firing position. The rifle should point naturally at the desired point. No muscular tension or movement should be necessary to hold the sights on target. To check the natural point of aim, the sniper assumes a comfortable, stable firing position. He then places his cheek on the stock at the correct stock weld, enters into the natural respiratory pause, looks away from the scope by moving only his eye, relaxes and lets the rifle drift to its natural point of aim, and then looks back into the scope. If the reticle is in the correct location on the target, the natural point of aim is correct. If it is not correct, the sniper must change his body position to bring the sights onto the target. If muscles are used to adjust the weapon onto the point of aim, the muscles will automatically relax as the rifle fires, and the rifle will begin to move toward its natural point of aim. Because this movement begins just before the weapon discharges, the rifle is moving as the bullet leaves the muzzle. This causes displaced shots with no apparent cause (recoil disguises the movement). By adjusting the weapon and body as a single unit, rechecking, and readjusting as necessary, the sniper achieves a true natural point of aim. Once the position is established, the sniper will then aim the weapon at the exact point on the target. Aiming involves three areas; eye relief, sight alignment, and sight picture.

(1) **Eye relief**. This is the distance from his firing eye to the rear sight or the rear of the scope tube. When using iron sights, the distance must remain consistent from shot to shot in any given firing position to preclude changing what the sniper views through the rear sight. Relief will, however, vary from firing position to firing position and from sniper to sniper according to their neck length, their angle of head approach to the stock, the depth of their shoulder pocket, and the firing position from which they are firing. This distance (Figure 3-2) is more rigidly controlled with telescopic sights than with iron sights.

He must take care to prevent eye injury caused by the scope tube striking his brow during recoil. Regardless of the sighting system he uses, he must place his head as upright as possible with his firing eye located directly behind the rear portion of the sighting system. This head placement also allows the muscles surrounding his eye to relax. Incorrect head placement causes the sniper to have to look out of the top or corner of his eye, resulting in muscular strain. Such strain leads to blurred vision and can also cause eye strain. Eye strain can and should be avoided by not staring through the iron or telescopic sights for extended periods. The best aid to consistent eye relief is maintaining the same stock weld from shot to shot.

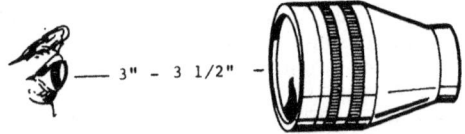

Figure 3-2. Eye relief.

(2) <u>Sight alignment</u>. With iron sights, this is the relationship between the front and rear sights as seen by the sniper (Figure 3-3). The sniper centers the top edge of the front sight blade horizontally and vertically within the rear aperture. (The center of aperture is easiest for the eye to locate and allows the sniper to be consistent in blade location.) With telescopic sights, sight alignment is the relationship between the cross hairs (reticle) and a full field of view as seen by the sniper. The sniper must place his head so that a full field of view completely fills the tube, with no dark shadows or crescents. This will cause misplaced shots. The sniper centers the reticle in a full field of view, ensuring the vertical cross hair is straight up

and down so the rifle is not canted. Again, the center is easiest for the sniper to locate and allows for consistent reticle placement.

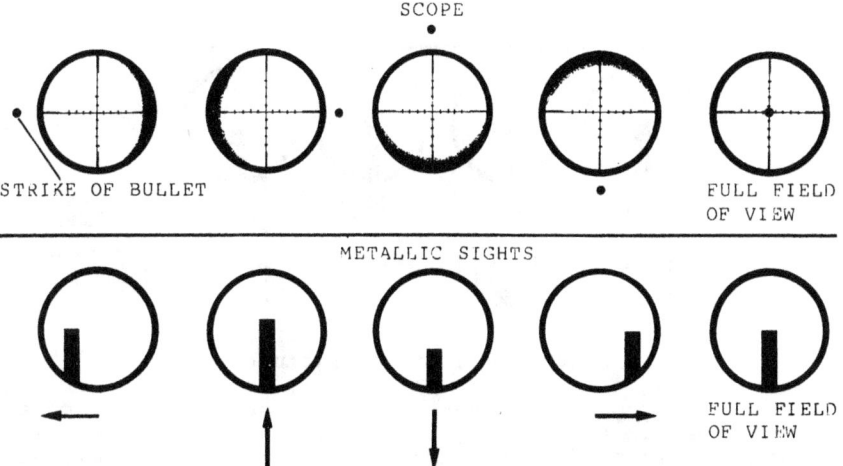

Figure 3-3. Sight alignment.

(3) **Sight picture**. With iron sights, this is the relationship between the rear aperture, the front blade, and the target as seen by the sniper (Figure 3-4). The sniper centers the top edge of the blade in the rear aperture. He then places the top edge of the blade in the center of the largest visible mass of the target (disregard the head and use the center of the torso). With telescopic sights, sight picture is the relationship between the reticle and full field of view and the target as seen by the sniper. He centers the reticle in a full field of view. He then places the reticle center of the largest visible mass of the target (as in iron sights). The center of mass of the target is easiest for the sniper to locate, and it surrounds the intended point of impact with a maximum amount of target area in case of an error in the aiming process.

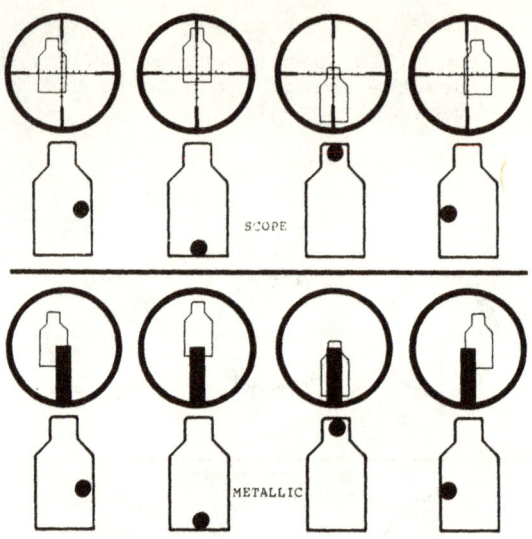

Figure 3-4. Sight picture.

(4) <u>Aiming process error</u>. Everyone occasionally makes an error in the aiming process. When alignment and picture are perfect (regardless of sighting system), and when everything else is done correctly, the shot will hit center of mass on the target. However, with an error in sight alignment, the bullet is displaced in the direction of the error. Such an error creates an angular displacement between the line of sight and the line of bore. This displacement increases as range increases. The amount of bullet displacement depends on the size of alignment error. Close targets will show little or no discernible error. Distant targets can show appreciable displacement or can be missed altogether with severe sight misalignment. Inexperienced snipers are prone to make this kind of error. They are unsure of what correctly aligned sights look like (especially telescopic sights); they vary their head positions (and eye relief) from shot to shot; and they are apt to make mistakes while firing.

(5) <u>Sight picture error</u>. An error in sight picture is an error in the placement of the aiming point. Such an error causes no displacement between the line of sight and the line of bore. The weapon is simply pointed at the wrong spot on the target. Because no displacement exists as range increases, close and far targets are hit

or missed depending on where the front sight or the reticle is when the rifle fires. All snipers face this kind of error every time they shoot. This is because, regardless of firing position stability, the weapon will always be moving. A supported rifle will move much less than an unsupported one, but both will still move in what is known as a wobble area. The sniper must adjust his firing position so that his wobble area is as small as possible and centered on the target. With proper adjustments, the sniper should be able to fire the shot while the front sight blade or reticle is on the target at, or very near, the desired aiming point. How far the blade or reticle is from this point when the weapon fires is the amount of sight picture error all snipers face.

(6) <u>Dominant eye</u>. Some individuals may have difficulty aiming because of interference from their dominant eye, if this is not the eye used in the aiming process. This may require the sniper to fire from the other side of the weapon (right-handed firer will fire left-handed). To determine which eye is dominant, hold an index finger 6 to 8 inches in front of your eyes. Close one eye at a time while looking at the finger; one eye will make the finger appear to move and the other will not. The eye that does not make the finger appear to move is the dominant eye.

c. Exercising Breath Control. Breath control is important with respect to the aiming process. If the shooter breathes while trying to aim, the rise and fall of his chest causes the rifle to move. He must, therefore, accomplish sight alignment during breathing and complete the aiming process while holding his breath. To do this, he first inhales then exhales normally and stops at the moment of natural respiratory pause.

(1) A respiratory cycle lasts 4 to 5 seconds. Inhalation and exhalation require only about 2 seconds. Thus, between each respiratory cycle there is a pause of 2 to 3 seconds. This pause can be extended to 10 seconds without any special effort or unpleasant sensations. The shooter should fire during this pause when his breathing muscles relax. This avoids strain on his diaphragm.

(2) A shooter should assume his firing position and breathe naturally until his hold begins to settle. Many shooters then take a slightly deeper breath, exhale, and

pause, expecting to fire the shot during the pause. If the hold does not settle sufficiently to allow the shot to be fired, the shooter resumes normal breathing and repeats the process.

(3) The respiratory pause should never feel unnatural. If it is too long, the body suffers from oxygen deficiency and sends out signals to resume breathing. These signals produce involuntary movements in the diaphragm and interfere with the shooter's ability to concentrate. Generally speaking, 8 to 10 seconds is the maximum safe period for the respiratory pause. During multiple, rapid engagements, the breathing cycle should be forced through a rapid, shallow cycle between shots instead of trying to hold the breath or breathing. Firing should be accomplished at the forced respiratory pause.

d. Exercising Trigger Control. Trigger control is the most important of the sniper marksmanship fundamentals. It is defined as causing the rifle to fire when the sight picture is at its very best, without causing the rifle to move. Trigger squeeze, on the other hand, is defined as the independent action of the forefinger on the trigger, with a uniformly increasing pressure straight to the rear until the rifle fires. Trigger control is the last task to be accomplished before the weapon fires. It is more difficult to apply when using a telescope or when a firing position becomes less stable.

(1) Proper trigger control occurs when the sniper places his firing finger as low on the trigger as possible and still clears the trigger guard, thereby achieving maximum mechanical advantage. He engages the trigger with that part of his firing finger that allows him to pull the trigger straight to the rear. In order to avoid transferring movement of the finger to the entire rifle, the sniper should see daylight between the trigger finger and the stock as he squeezes the trigger straight to the rear. He fires the weapon when the front blade or reticle is in a position to ensure a well-placed shot.

(2) As the stability of a firing position decreases, the wobble area increases. The larger the wobble area, the harder it is to fire the shot without reacting to it. This reaction occurs when the sniper --

(a) Anticipates recoil. The firing shoulder begins to move forward just before the round fires.

(b) **Jerks the trigger.** The trigger finger moves the trigger in a quick, choppy, spasmodic attempt to fire the shot before the front blade or reticle can move away from the desired point of aim.

(c) **Flinches.** The sniper's entire upper body (or parts thereof) overreacts to anticipated noise or recoil. This is usually due to being unfamiliar with the weapon.

(d) **Avoids recoil.** The sniper tries to avoid recoil or noise by moving away from the weapon, or closing the firing eye, just before the round fires. This, again, is caused by a lack of knowledge of the weapon's actions upon firing.

(3) Trigger control is best handled by assuming a stable position, adjusting on the target, and beginning a breathing cycle. As the sniper exhales the final breath toward a natural respiratory pause, he secures his finger on the trigger. On the M21, he will take up the slack in the trigger until resistance is felt. As the front blade or reticle settles at the desired point of aim, and the natural respiratory pause is entered, he applies initial pressure. He increases the tension on the trigger during the respiratory pause as long as the front blade or reticle remains in the area of the target that ensures a well-placed shot. If the front blade or reticle moves away from the desired point of aim on the target, and the pause is free of strain or tension, the sniper stops increasing the tension on the trigger, waits for the front blade or reticle to return to the desired point, and then continues to squeeze the trigger. This is trigger control. If movement is too large for recovery or if the pause has become uncomfortable (extended too long), the sniper should, whenever possible, carefully release the pressure on the trigger and begin the respiratory cycle again.

3-2. FOLLOWING THROUGH

Applying the fundamentals increases the odds of a well-aimed shot being fired. There are, however, additional skills that, when mastered, make that first-round kill even more of a certainty. One of these is to follow through.

a. This is the act of continuing to apply all the sniper marksmanship fundamentals as the weapon fires as well as immediately after it fires. Follow-through consists of --

o Keeping the head in firm contact with the stock (stock weld).

o Keeping the finger on the trigger all the way to the rear.

o Continuing to look through the rear aperture or scope tube.

o Ensuring that muscles stay relaxed.

o Avoiding reacting to recoil and or noise.

o Releasing the trigger only after the recoil has stopped.

b. Good follow-through ensures the weapon is allowed to fire and recoil naturally, and the sniper-rifle combination reacts as a single unit to such actions.

3-3. CALLING THE SHOT

Calling the shot is being able to tell where the round should impact on the target. Because live targets invariably move when hit, the sniper will find it difficult, if not impossible, to use his scope to locate the target after the round is fired. Using iron sights, the sniper will find that searching for a downrange hit is beyond his capabilities. The sniper must be able to accurately call his shots. Proper follow-through will aid in calling the shot. The dominant factor in shot calling is, however, where the reticle or post is located when the weapon discharges. This location is called his final focus point.

a. With iron sights, the final focus point should be on the top edge of the front sight blade. The blade is the only part of the sight picture that is moving (in the wobble area). Focusing on it aids in calling the shot and detecting any errors in sight alignment or sight picture. Of course, lining up the sights and the target initially requires the sniper to shift his focus from the target to the blade and back until he is satisfied that he is properly aligned with the target. This shifting exposes two more facts about eye focus. The eye is capable of instantly shifting focus from near objects (the blade) to far objects (the target). The eye cannot, however, be focused so that two objects at greatly different ranges (again the blade and target) are both in sharp focus. After years of experience, many snipers find that they no longer hold final focus on the front sight blade. Their focus is somewhere between the blade and the target. This act has been related to many things, from personal preference to failing eyesight. Regardless, inexperienced snipers are still advised to use the blade as a final focus point.

b. The final focus is easily placed with telescopic sights because of their optical qualities. Properly focused, a scope should present both the field of view and the reticle in sharp detail. Final focus should then be on the target. While focusing on the target, if the head is moved slightly from side to side, the reticle may seem to move across the target face, even though the rifle and scope are virtually motionless. This movement is parallax. Parallax is present when the target image is not correctly focused on the reticle's focal plane. Therefore, the target image and the reticle will appear to be in two separate positions inside the scope, causing the effect of reticle movement across the target. A small amount of parallax will be unavoidable throughout the range of the ART series of scopes. The M3A scope on the M24 has a focus adjustment that eliminates parallax in the scope. The sniper should adjust the focus knob until the target's image is on the same focal plane as the reticle. To determine if the target's image appears at the ideal location, the sniper should move his head slightly left and right to see if the reticle appears to move. If it does not move, the focus is properly adjusted and no parallax will be present.

3-4. ZEROING THE RIFLE

When a sniper fires a shot that does not strike the desired location on the target, he must move the sights to move the shot to that point. The zero of a rifle is that elevation and windage setting required to place a shot or shot group at a given point, at a given range, on a day when no wind is blowing. Experience has shown that the best way to zero a rifle is to shoot from the desired firing position, at the desired range, and with the intended cadence (rate of fire).

a. Depending on the situation, a sniper may be called upon to deliver an effective shot at ranges up to 900 meters. This requires that he zero his rifle at most of the ranges that he may be expected to fire. When using the ART series of scopes, he should zero at 300 meters and confirm the zero at the more distant targets. When using the M3A scope, he should zero at 100 meters and then confirm the zero at the more distant targets. His success depends on a one-round, one-kill philosophy. He may not get the second shot for obvious reasons. Therefore, he must zero his rifle so accurately that when he applies the fundamentals, he can be assured of a definite kill.

    b. Once a sniper obtains a zero, he cannot expect the zero to remain absolutely constant. He must periodically confirm the zero, such as after disassembly of the sniper rifle for maintenance and cleaning, for changes in ammunition lots, for changes in altitude, or as the result of severe weather changes. These changes in zero and the conditions that caused them should be recorded in the sniper's data book (see Appendix C). The individual who will use the rifle must zero it. Individual characteristics such as stock weld, eye relief, firing position, and trigger control usually result in each sniper having a different zero with the same rifle, or a change in zero from one firing position to another. It is important that the sniper knows his first round zero, especially at longer ranges. Some rifles will place the first round out of the main group. By keeping written data over a period of time, the sniper knows what his rifle will do when the barrel is cold and clean, or cold and fouled. First shot variance is more likely to occur from a cold, clean barrel, especially if the barrel is not absolutely dry.

3-5. CONSIDERING WEATHER EFFECTS

In the case of the highly trained sniper, effects of the weather are a primary cause of error in the strike of the bullet. Wind, mirages, light, temperature, and humidity affect the bullet, the sniper, or both. Some effects are insignificant; however, sniping is often done in extremes of weather; therefore, all effects must be considered.

    a. Wind poses the biggest problem to the sniper. The effect that wind has on the bullet increases with range. This is due primarily to the slowing of the bullet's velocity combined with a longer flight time. This allows the wind to have a greater effect on the round as distances increase. The result is a loss of stability. Wind also has a considerable effect on the sniper. The stronger the wind, the more difficult it is to hold the rifle steady. This can be partially offset with training, conditioning, and by the use of supported firing positions. (See Appendix E for a windage conversion table.)

    b. Before making any sight adjustment to compensate for wind, the sniper must determine its direction and velocity. He may use certain indicators to accomplish this. These are range flags, smoke, trees, grass, rain, and the sense of feel. However, the most preferred method of determining wind direction and velocity is reading mirage (see paragraph 3-5e). In most cases, it is relatively easy to determine the direction the wind is blowing simply by observing the indicators.

(1) A common method of estimating the velocity of the wind during training is to watch the range flag (Figure 3-5). The sniper determines the angle between the flag and pole, in degrees, then divides by the constant number 4. The result gives the approximate velocity in miles per hour.

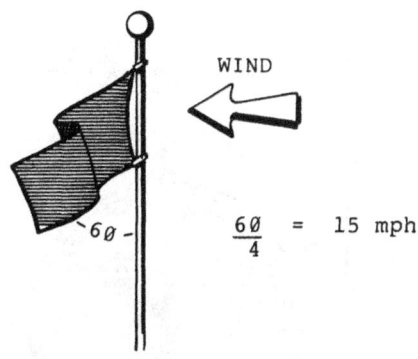

Figure 3-5. The flag method.

(2) If no flag is visible, the sniper holds a piece of paper, grass, cotton, or some other light material at shoulder level, then drops it. He then points directly at the spot where it lands and divides the angle between his body and arm by the constant number 4. This gives him the approximate wind velocity in miles per hour.

(3) If for some reason these methods cannot be used, the following information is helpful in determining velocity. Winds under 3 miles per hour can barely be felt, but they may be determined by smoke drifts. A 3- to 5-mile-per-hour wind can barely be felt on the face. With a 5- to 8-mile-per-hour wind, the leaves in the

3-13

trees are in constant motion, and with a 12- to 15-mile-per-hour wind, small trees begin to sway.

c. Since the sniper must know how much effect the wind will have on the bullet, he must be able to classify the wind. The best method is to use the clock system (Figure 3-6). With the sniper at the center of the clock and the target at 12 o'clock, the wind is assigned three values: full, half, and no value. Full value means that the force of the wind will have a full effect on the flight of the bullet. These winds come from 3 and 9 o'clock. Half value means that a wind at the same speed, but from, 1, 2, 4, 5, 7, 8, 10, and 11 o'clock, will move the bullet only half of the full value wind. No value means that a wind from 6 or 12 o'clock will have little or no effect on the flight of the bullet.

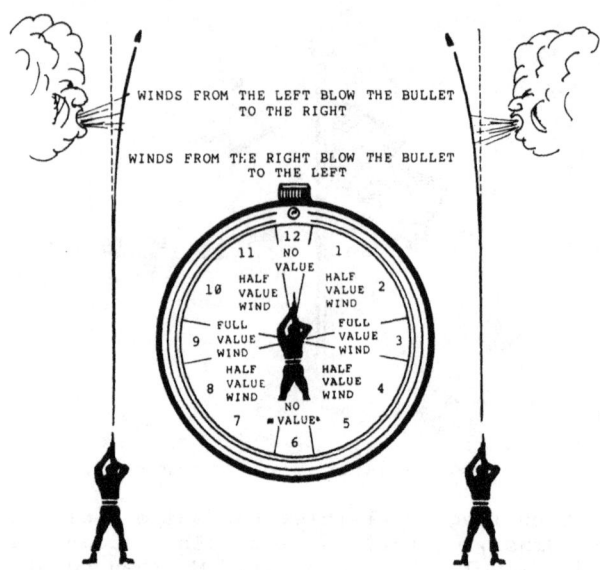

Figure 3-6. Clock system.

d. After determining wind direction and velocity, the sniper bases the windage correction on the following formula:

$$\frac{(R\ 100)\ \times\ V}{constant}$$ number of minutes of angle for a full value wind on a sniper rifle.

3-14

In this formula, R is the range in hundreds of meters and V is the velocity of the wind in miles per hour. For half value winds, divide the answer by 2.

The constant for the M118 special ball round is 20. This constant may not remain correct if other ammunition is used. Changes in bullet weight and velocity affect performance characteristics.

EXAMPLE: The wind is blowing from 9 o'clock at 5 miles per hour. The range is 800 meters; using the wind formula, R = 8 and V = 5, and the constant 20 for 800 meters, the correction is:

$$\frac{(R \div 100) \times V}{20} = \frac{8 \times 5}{20} = 2 \text{ MOA}.$$

e. A mirage is a reflection of the heat through layers of air at different temperatures and density as seen on a warm day (Figure 3-7). With the telescope, a mirage can be seen as long as there is a difference in ground and air temperatures. Proper reading of the mirage will enable the sniper to estimate and make windage corrections with a high degree of accuracy.

f. As observed through the telescope, the mirage appears to move with the same velocity as the effective wind, except when blowing straight into or away from the scope. Then the mirage gives the appearance of moving straight up with no lateral movement. This is termed a boiling mirage. In general, changes in the velocity of the wind, up to approximately 12 miles per hour, can be readily determined by observing the mirage. Beyond that speed, the movement of the mirage is too fast for detection of minor changes.

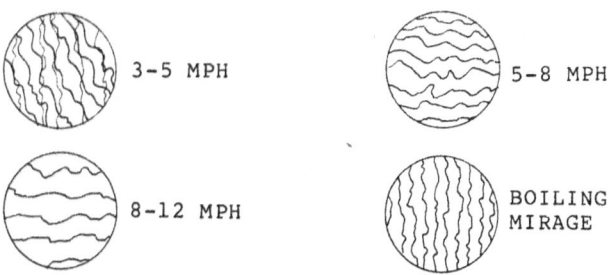

Figure 3-7. Types of mirages.

3-15

g. Temperature affects the elevation setting required to hit the center of the target (Figure 3-8). This is because an increase in temperature of 20 degrees Fahrenheit increases the muzzle velocity by approximately 50 feet per second. Regardless of the range, the sniper must change his elevation adjustments about one minute for each 20-degree change in temperature. For a drop in temperature, he raises the elevation; for an increase in temperature, he lowers the elevation.

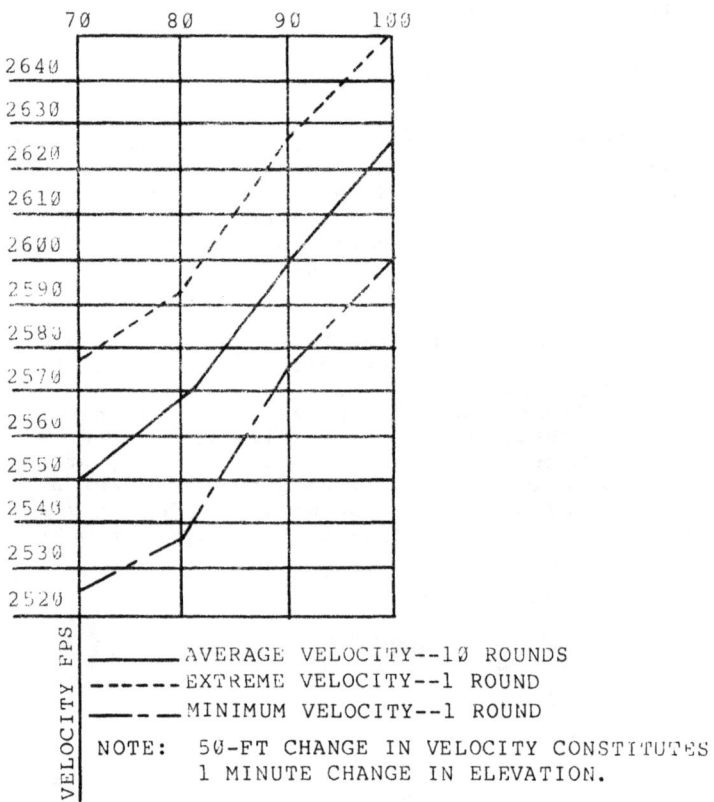

Figure 3-8. Temperature effects.

h. Light may or may not affect the sniper's aim; it affects different people in different ways. Light has a greater effect when shooting with iron sights. Telescopic sights will be slightly affected, if at all. The general tendency, however, is for the sniper to shoot high on a dull, cloudy day and low on a bright, clear day. Extreme light conditions from the left or the right may affect the horizontal impact of a shot or shot group.

i. To understand the effects of humidity on the strike of the bullet, one must realize that the higher the humidity, the denser the air. Therefore, there is more resistance to the flight of the bullet. Resistance tends to slow the bullet and as a result, the sniper must raise his elevation to compensate for it. The effect of humidity at short ranges is not as great as at long ranges. Again, the experience of the sniper and his resultant study of hits and groups under varied conditions of humidity will determine the effects of humidity on his zero.

j. By not considering all of the effects of weather, some snipers tend to overemphasize certain effects. This will produce bad shots from time to time. As previously mentioned, snipers normally fire for a certain period of time under average conditions. As a result they zero their rifle, and (with the exception of minor displacements of shots and groups) they do not have difficulty except for the wind. Yet a sniper can travel to a different location to fire again and find a change in zero. A thorough study of the weather effects would indicate the change. Proper recording and study based on experience are all-important with respect to determining the effects of weather. Probably one of the most difficult things to impress upon a sniper is the evidence of a probable change in his zero. If a change is indicated, it applies at all ranges.

3-6. FIRING ONE ROUND

Once the sniper has been taught the fundamentals, his primary concern in ensuring the quality of his shooting is his ability to apply this knowledge during his mission. A very effective way to do this is to teach the sniper team the integrated act of firing one round. Firing one round involves the preparation of equipment, the selection of a firing position, the detection of a target, and the sight adjustments needed to engage the target.

a. Before leaving the preparation area, the sniper ensures his weapon system and equipment are in working condition. The weapon should be properly camouflaged, zeroed with the ammunition he will use on the mission, and have a bore that is clean and dry. The observer will inspect the binoculars and telescope to ensure the optics are clean and that they will not fog up on the mission.

b. The sniper team carefully selects a firing position, ensuring it has enough room for the sniper and observer to acquire good firing and observing positions with clear fields of fire. The sniper should be in a prone supported position that will allow him to fire in any direction within the target area. Once in position, the sniper team will conduct a search of the area while preparing a range card to use for quick range referencing.

c. Upon detection of a target, the team determines the distance to the target, and the sniper makes the needed elevation adjustment. The observer determines the wind direction and velocity and tells the sniper the needed windage adjustment. Using good marksmanship fundamentals, the sniper fires one round. The observer watches the target and bullet trace to determine the exact placement of the round and prepares to give corrections to the sniper, if necessary.

3-7. HOLDING OFF FOR ELEVATION OR WIND

This technique is used only when the sniper does not have time to change his sight setting. The sniper rarely achieves pinpoint accuracy when holding off, since a minor error in range determination or a lack of a precise aiming point might cause the bullet to miss the desired point. He uses holdoff with the sniperscope only if several targets appear at various ranges, and time does not permit adjusting the scope for each target.

a. Holdoff is used to hit a target at ranges other than the range for which the rifle is presently adjusted. When aiming directly at a target at ranges greater than the set range, the bullet will hit below the point of aim. At lesser ranges, the bullet will hit higher than the point of aim. If the sniper understands this and knows about trajectory and bullet drop, he will be able to hit the target at ranges other than that for which the rifle was adjusted. For example, the rifle is adjusted for a target located 500 meters downrange and another target appears at a range of 600 meters. The holdoff would be 25 inches;

that is, the sniper should hold off 25 inches above the center of visible mass in order to hit the center of mass of that particular target. If another target were to appear at 400 meters, the sniper would aim 15 inches below the center of visible mass in order to hit the center of mass. With the M3A scope, the vertical mil dots on the scope's reticle can be used as aiming points when using elevation holdoffs. For example, if the sniper has to engage a target at 500 meters and the scope is set at 400 meters, he would place the first mil dot 5 inches below the vertical line on the target's center mass. This will give the sniper a 15-inch holdoff at 500 meters.

b. The sniper may use holdoff in the following ways to compensate for the effect of wind:

(1) When using the horizontal stadia marks on the ART-type scope reticle to measure the required holdoff distance, the sniper must remember to first range-in on the target. He then subdivides the horizontal reticle line within the stadia marks (60 inches) to determine the correct distance for holdoff. He can also use that reference point as an aiming point or point of aim. When using the M3A scope, the sniper uses the horizontal mil dots on the reticle to hold off for wind. For example, if the sniper has a target at 500 meters that requires a 10-inch holdoff, the sniper would place the target's center mass halfway between the cross hair and the first mil dot (1/2 mil).

(2) When holding off, the sniper aims into the wind. If the wind is from the right, his point of aim is to the right. If the wind is from the left, his point of aim is to the left.

(3) Constant practice in wind estimation can bring about proficiency in making sight adjustments or learning to hold off correctly. If the sniper misses the target and the impact of the round is observed, the sniper notes the lateral distance of his error and refires, holding off that distance in the opposite direction.

3-8. ENGAGING MOVING TARGETS

Engaging moving targets not only requires the sniper to determine the target distance and wind effects on the round, but he must also consider the lateral speed of the target, the round's time of flight, and the placement of a proper lead to compensate for both. These added variables increase the possibility of a miss. Therefore, the sniper should engage moving targets when it is the only option.

a. Leading. Engaging moving targets requires the sniper to place the cross hairs ahead of the target's movement. The distance the cross hairs are placed in front of the target's movement is called a lead. There are four factors in determining leads.

(1) Speed of the target. As a target moves faster, it will move a greater distance during the bullet's flight. Therefore, the lead will increase as the target's speed increases.

(2) Angle of movement. A target moving perpendicular to the bullet's flight path will move a greater lateral distance during its flight time than a target moving at an angle away from or toward the bullet's path. Therefore, a target moving at a 45-degree angle will have less lateral movement than a target moving at a 90-degree angle. As the lateral movement increases, the lead must be increased.

(3) Range to the target. The farther away a target is, the longer it will take for the bullet to reach it. Therefore, the lead must be increased as the distance to the target increases.

(4) Wind effects. The sniper must consider how the wind will affect the trajectory of the round. A wind blowing opposite to the target's direction will require more of a lead than a wind blowing in the same direction as the target's movement.

b. Tracking. Tracking requires the sniper to establish an aiming point ahead of the target's movement and maintain it as the weapon is fired. This requires the weapon and body position to be moved while following the target and firing.

c. Trapping. Trapping or ambushing is the sniper's preferred method of engaging moving targets. This requires the sniper to establish an aiming point ahead of the target and pull the trigger when the target reaches it. This method allows the sniper's weapon and body position to remain motionless. With practice, snipers can determine exact leads and aiming points using the horizontal stadia lines in the ART scopes or the mil dots in the M3A.

d. **Preventing Errors.** When engaging moving targets, common errors are usually made because the sniper is under greater stress than with a stationary target. There are more considerations, such as retaining a steady position and the correct aiming point, how fast the target is moving, and how far away it is. The more practice a sniper has shooting moving targets, the better he will become. Some common mistakes are:

(1) The sniper has a tendency to watch his target instead of his aiming point. He must force himself to watch his lead point.

(2) The sniper may jerk or flinch at the moment his weapon fires because he thinks he must fire right now. This can be overcome through practice on a live-fire range.

(3) The sniper may hurry and thus forget to apply wind as needed. Windage must be calculated for moving targets just as for stationary targets. Failure to do this when acquiring a lead will result in a miss.

e. **Determining the Lead.** Once the required lead has been determined, the sniper should use the mil scale in the scope for precise holdoff. The mil scale can be mentally sectioned into 1/4-mil increments for leads. The chosen point on the mil scale becomes the sniper's point of concentration just as the cross hairs are for stationary targets. The sniper concentrates on the lead point and fires the weapon when the target is at this point. The following formulas are used to determine moving target leads:

TIME OF FLIGHT X TARGET SPEED = LEAD.

   Time of flight = flight time of the round in seconds.

   Target speed = speed the target is moving in feet per second.

   Lead = distance aiming point must be placed ahead of movement in feet.

   Average speed of a man during--

```
    Slow patrol = 1 fps/0.8 mph
    Fast patrol = 2 fps/1.3 mph
    Slow walk   = 4 fps/2.5 mph
    Fast walk   = 6 fps/3.7 mph
```

To convert leads in feet to meters:

LEAD IN FEET X 0.3048 = METERS

To convert leads in meters to mils:

$$\frac{\text{LEAD IN METERS} \times 1,000}{\text{RANGE TO TARGET}} = \text{MIL LEAD}$$

TC 23-14

CHAPTER 4
FIELD TECHNIQUES

The general mission of the sniper is to reduce selected enemy targets with long-range precision fire. How well he accomplishes this mission depends on his knowledge, understanding, and application of the various field techniques or skills that allow him to move, hide, observe, and detect his targets. This chapter discusses the field techniques and skills that the sniper must learn before his employment in support of combat operations. The sniper's application of these skills will affect his survival on the battlefield.

4-1. CAMOUFLAGE

Camouflage is one of the basic weapons of war. It can mean the difference between a successful or unsuccessful mission. To the sniper, it can mean the difference between life and death. Camouflage measures are very important since the sniper cannot afford to be detected at any time while moving alone, as part of another element, or while operating from a firing position. Marksmanship training teaches the sniper to hit a target, and a knowledge of camouflage teaches him how to escape becoming a target himself. Paying close attention to camouflage fundamentals is a mark of a well-trained sniper. (See FM 5-20 for more details.)

a. To become proficient in camouflage, the sniper must first understand target indicators. Target indicators are anything a person does or fails to do that could result in being detected. A sniper must know and understand target indicators to not only move undetected, but to detect enemy movement. Target indicators are sound, movement, improper camouflage, disturbance of wildlife, and odors.

(1) Sound.

o Most prominent during hours of darkness.

o Caused by movement, equipment rattling, or talking.

o Small noises may be dismissed as natural, but talking will not.

(2) Movement.

o Most prominent during hours of daylight.

o The human eye is attracted to movement.

4-1

o   Quick or jerky movement will be detected faster than slow movement.

(3) <u>Improper camouflage</u>.

o   Shine.

o   Outline.

o   Contrast with the background.

(4) <u>Disturbance of wildlife</u>.

o   Birds suddenly flying away.

o   Sudden stop of animal noises.

o   Animals being spooked.

(5) <u>Odors</u>.

o   Cooking.

o   Smoking.

o   Soap, lotions.

o   Insect repellents.

b. There are three fundamental methods the sniper can use to camouflage himself or his position. The sniper may use just one of these methods or a combination of all three to accomplish a given mission. The three methods are:

(1) <u>Hiding</u>. Hiding is completely concealing the body from observation by lying behind an object or thick vegetation.

(2) <u>Blending</u>. Blending is achieved by skillfully matching personal camouflage with the surrounding area to a point where the sniper is indiscernible.

(3) <u>Deceiving</u>. Deceiving is a technique used to trick the enemy into false conclusions about the location of the sniper.

c. There are two types of camouflage that the sniper can use to camouflage himself and his equipment. The two types are:

(1) **Natural**. Natural camouflage is vegetation or materials that are native to the given area. The sniper should always augment his appearance by using some natural camouflage.

(2) **Artificial**. Artificial camouflage is any material or substance that is produced for the purpose of coloring or covering something in order to conceal it. Examples are:

    (a) Camouflage sticks and face paint. Camouflage sticks or face paints are used to cover all exposed areas of skin, such as face, hands, and the back of the neck. The parts of the face that form shadows should be lightened and the parts that shine should be darkened. There are three types of camouflage patterns used by the sniper. They are:

       o Striping - used when in heavily wooded areas, and leafy vegetation is scarce.

       o Blotching - used when area is thick with leafy vegetation.

       o Combination - used when moving through changing terrain. It is normally the best all-around pattern.

    (b) Ghillie suit. The term "ghillie suit" originated in Scotland during the 1800s. Scottish game wardens made special camouflage suits in order to catch poachers. The ghillie suit today is a specially made camouflage uniform that is covered with irregular patterns of garnish or netting (Figure 4-1). Ghillie suits can be made from BDUs or one-piece aviator type uniforms. Turning the uniform inside out will place the pockets inside the suit. This protects items in the pockets from damage caused by crawling on the ground. The front of the ghillie suit should be covered with canvas or some type of heavy cloth to reinforce it. The knees and elbows should be covered with two layers of canvas since these areas are prone to wear out more often. The garnish or netting should cover the shoulders and reach down to the elbows on the sleeves. The garnish applied to the back of the suit should be long enough to cover the sides of the sniper when he is in the prone position. A bush hat is also covered with garnish or netting. The garnish should be long enough to break up the outline of the sniper's neck, but not long enough in front to obscure his vision or hinder movement. A veil can be made from a net or piece of cloth covered with garnish or netting. It is used to cover the weapon and sniper's head when in a firing position. The veil can be sewn into the ghillie suit or carried separately. Remember, a ghillie suit does not make one invisible and is only a camouflage base. Natural vegetation should be added to help blend with the surroundings.

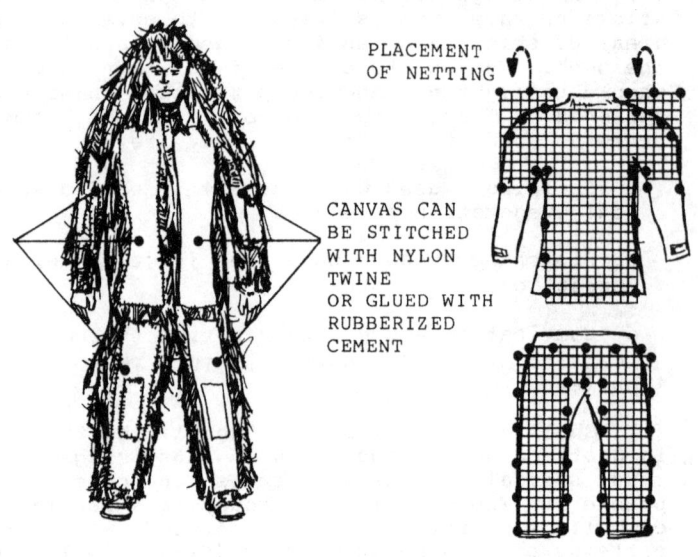

Figure 4-1. Ghillie suit.

(c) Field expedient camouflage. The sniper may have to use field expedient camouflage if the earlier mentioned methods are not available. Instead of camouflage sticks or face paint, the sniper may use charcoal, walnut stain, mud, or whatever will work. He will not use oil or grease because of the strong odor that it gives off. Natural vegetation can be attached to the body by boot bands or rubber bands, or by cutting holes in the uniform.

(d) Equipment camouflage. The sniper must also camouflage all equipment he will use. However, he will ensure the camouflage does not interfere with or hinder the operation of the equipment.

o Rifles. The sniper weapon system and the M16 should also be camouflaged to break up their outlines. He will not bind the scope of the M21 to a point that it will not properly adjust or have loose garnish that will get caught in the bolts of the rifles. The sniper weapon system can be carried in a "drag bag" (Figure 4-2), which is a rifle case made of canvas and covered with garnish similar to the ghillie suit. However, the rifle will not be combat ready while it is in the drag bag.

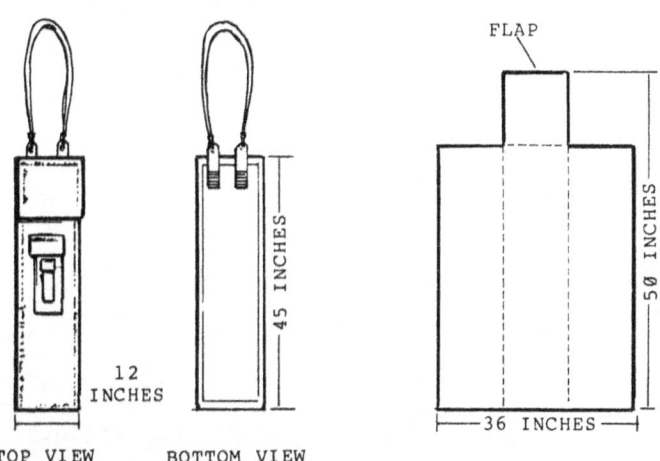

Figure 4-2. Drag bag.

o Optics. Optics used by the sniper must also be camouflaged to break up the outline and reduce the possibility of light reflecting off the lenses. Lenses can be covered with mesh type webbing or nylon hose material.

o ALICE pack. If the ALICE pack is to be used while wearing the ghillie suit, it must be camouflaged in the same manner as the suit.

(e) Geographic camouflage. The sniper must alter his camouflage to blend in with various changes of vegetation and terrain in different geographic areas. Examples of such changes are:

o Snow areas. Blending of colors is more effective than texture camouflage in snowy areas. In areas with heavy snow or in wooded areas with trees covered with snow, a full white camouflage suit should be worn. In areas with snow on the ground but not on the trees, white trousers with green and brown tops should be worn.

o Desert areas. In sandy desert areas that have little vegetation, the blending of tan and brown colors is important. In these areas, the sniper must make full use of the terrain and the vegetation that is available to remain unnoticed.

o Jungle areas. In jungle areas, texture camouflage and contrasting colors must be used. Natural vegetation must also be used.

o Urban areas. In urban areas, the sniper's camouflage should be a blended color (shades of gray usually work best). Texture camouflage is not as important in these environments.

The sniper must be camouflage conscious from the time he departs on a mission until he returns. He must constantly use the terrain, vegetation, and shadows to remain undetected. At no other time during the mission will the sniper have a greater tendency to be careless than during his return to a friendly area. Fatigue and undue haste may override caution and planning. Therefore, the sniper needs to pay particular attention to his camouflage discipline on his return from missions.

4-2. COVER AND CONCEALMENT

The proper understanding and application of the principles of cover and concealment used with the proper application of camouflage will protect the sniper from enemy observation.

a. Cover is natural or artificial protection from the fire of enemy weapons. Natural cover (ravines, hollows, reverse slopes) and artificial cover (fighting positions, trenches, walls) protect the sniper from flat trajectory fires and partially protect him from high-angle fires and the effects of nuclear explosions. Even the smallest depression or fold in the ground may provide some cover when the sniper needs it most. A 6-inch depression, properly used, may provide enough cover to save his life under fire. The sniper must always look for and take advantage of every bit of cover the terrain offers. By combining this habit with proper movement techniques, he can protect himself from enemy fire. To get protection from enemy fire when moving, he uses routes that put cover between him and the places where the enemy is known or thought to be. He uses natural and artificial cover to keep the enemy from seeing and firing at him.

b. Concealment is natural or artificial protection from enemy observation. The surroundings may provide natural concealment that needs no change prior to use (bushes, grass, and shadows). The sniper creates artificial concealment from materials such as burlap and camouflage nets, or he can move natural materials (bushes, leaves, and grass) from their original location. The sniper must consider the effects of the change of seasons on the concealment provided by both natural and artificial materials. The principles of concealment include the following:

(1) Avoid unnecessary movement. Remain still -- movement attracts attention. The sniper's position may be concealed when he remains still, yet easily detected if he moves. His movement against a stationary background makes him stand out clearly. When he must change positions, he moves carefully over a concealed route to the new position, preferably during limited visibility. He moves inches at a time, slowly and cautiously, always scanning ahead for the next position.

(2) Use all available concealment.

(a) Background. Background is important; the sniper must blend with it to prevent detection. The trees, bushes, grass, earth, and man-made structures that form the background vary in color and appearance. This makes it possible for the sniper to blend with them. He selects trees or bushes that blend with his uniform and absorb the outline of his figure. He must always assume that his area is under observation.

(b) Shadows. The sniper in the open stands out clearly, but the sniper in the shadows is difficult to see. Shadows exist under most conditions, day and night. A sniper should never fire from the edge of a wood line; he should fire from a position inside the wood line (in the shade or shadows provided by the tree tops).

(3) <u>Stay low to observe</u>. A low silhouette makes it difficult for the enemy to see a sniper. Therefore, he observes from a crouch, a squat, or a prone position.

(4) <u>Expose nothing that shines</u>. Reflection of light on a shiny surface instantly attracts attention and can be seen from great distances. The sniper uncovers his rifle scope only when indexing and reducing a target. He uses optics cautiously in bright sunshine because of the reflections they cause.

(5) <u>Avoid skylining</u>. Figures on the skyline can be seen from a great distance, even at night, because a dark outline stands out against the lighter sky. The silhouette formed by the body makes a good target.

(6) <u>Alter familiar outlines</u>. Military equipment and the human body are familiar outlines to the enemy. The sniper alters or disguises these revealing shapes by using the ghillie suit or outer smock that is covered with irregular patterns of garnish. The sniper must alter his outline from his head to the soles of his boots.

(7) <u>Keep quiet</u>. Noise, such as talking, can be picked up by enemy patrols or listening posts. The sniper silences gear before a mission so that it makes no sound when he walks or runs.

4-3. MOVEMENT AND LAND NAVIGATION

A sniper team's mission and method of employment differ in many ways from those of the infantry squad. One of the most noticeable differences is the movement technique used by the snipers. Due to the nature of their mission, movement by snipers must not be detected or even suspected by the enemy. Because of this, a sniper must master individual sniper movement techniques.

   a. When moving, the sniper should always remember the following rules:

     (1) Always assume your area is under enemy observation.

     (2) Move slowly. A sniper counts his movement progress by feet and inches.

(3) Do not cause the overhead movement of trees, bushes, or tall grasses by rubbing against them.

(4) Plan every movement and move in segments of the route at a time.

(5) Stop, look, and listen often.

(6) Move during disturbances such as gunfire, explosions, aircraft noise, wind, or anything that will distract the enemy's attention or conceal the sniper's movement.

b. The individual movement techniques used by the sniper are designed to allow him to move with the least possibility of being detected. These movement techniques are:

(1) <u>Sniper low crawl</u>. This technique (Figure 4-3) is used when concealment is extremely limited, when in close proximity to the enemy, or when occupying a firing position. The sniper low crawl is conducted by doing the following:

(a) Lie flat on the ground with head turned to one side, arms in a straight line forward of the head, and legs together in a straight line to the rear with heels touching the ground.

(b) Hold the weapon by its sling and keep it in a straight line parallel to the body.

(c) Move by pushing slowly and evenly with the toes while pulling with the fingers, 2 to 4 inches at a time.

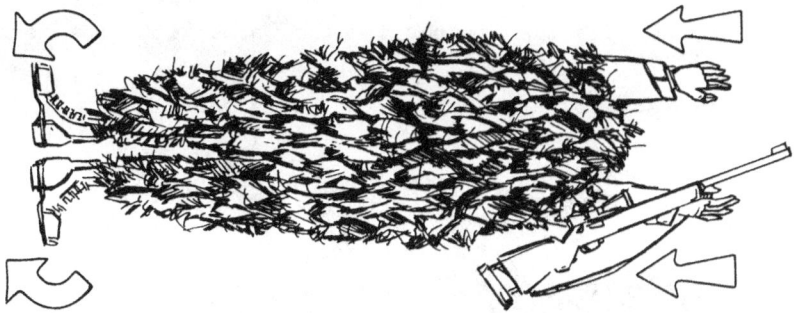

Figure 4-3. Sniper low crawl.

(2) <u>Medium crawl</u>. This technique (Figure 4-4) is used when concealment is limited and the sniper needs to move

faster than the sniper low crawl allows. The medium crawl is similar to the infantryman's low crawl and is executed by doing the following:

(a) Lie flat on the ground with head turned to one side, arms forward of the head, and legs spread apart to the rear.

(b) Hold the weapon by the forward sling swivel and lay it on top of the forearm.

(c) Move by pulling one leg forward and pushing while pulling with the arms, 12 to 16 inches at a time.

Figure 4-4. Medium crawl.

(3) <u>High crawl</u>. This technique (Figure 4-5) is used when concealment is limited but high enough to allow the sniper to raise his body off the ground. The high crawl is similar to the infantry high crawl and is executed by doing the following:

(a) Support the body by the elbows and knees.

(b) Cradle the weapon in the arms with the scope between the chin and chest.

(c) Move by alternating the knees and elbows.

Figure 4-5. High crawl.

(4) <u>Hand and knees crawl</u>. This technique (Figure 4-6) is used when some concealment is available and the sniper needs to move faster than the medium crawl allows. The hand and knees crawl is conducted by doing the following:

(a) Support the body with knees and one hand while cradling the weapon with the other arm by placing the scope in the armpit.

(b) Move by moving one knee at a time forward, then moving the arm forward.

Figure 4-6. Hand and knees crawl.

(5) <u>Walking</u>. This technique (Figure 4-7) is used when there is good concealment, it is not likely the enemy is in the vicinity, and speed is necessary. The walk is conducted by doing the following:

(a) Crouch with upper body bent forward and knees slightly bent.

(b) Carry the weapon on line with the body by grasping the forward sling swivel. Keep the muzzle pointed down.

(c) Move a step at a time. Plan the placement of every step by choosing a spot to place your foot before you move it from the previous spot.

Figure 4-7. Walking.

c. Snipers are employed in two-man teams consisting of one observer and one sniper. Normally, the observer will carry an M16/M203, the sniper will carry the sniper weapon system, and both will have sidearms. Because of this lack of personnel and firepower, the sniper team cannot afford to be detected by the enemy, nor will they be able to engage the enemy in sustained engagements.

d. Whenever possible, the sniper team should be attached to a security element (squad/platoon). The security element allows the team to reach its area of operations quicker and safer than can be expected by the team operating alone. Plus, the security element provides the team a reaction force should the team be detected. Snipers will use the following guidelines when attached to a security element:

   (1) The security element leader is in charge of the team while the team is attached to the element.

(2) Snipers will always appear as an integral part of the element.

(3) Snipers will wear the same uniform as the element members.

(4) Snipers will maintain proper intervals and positions in all formations.

(5) The sniper weapon system will be carried in line and close to the body, hiding its outline and barrel length.

(6) All equipment that is unique to snipers will be concealed from view (optics, ghillie suits, and so forth).

e. Once in the area of operation, the sniper team will separate from the security element and operate alone. Two examples of sniper teams separating from security elements are:

(1) Security element will provide security while snipers prepare for operation.

o Snipers don their ghillie suits and camouflage themselves and their equipment (if mission requires).

o Snipers ensure all equipment is secure and any nonessential equipment is cached (if mission requires).

o Once the team is prepared, it will assume a concealed position within the area, and then the security element will depart the area.

o Once the security element has left the area, the team will wait in position long enough to ensure that the team or the security element has not been compromised, then move on to its tentative position.

(2) The security element conducts a short security halt at the separation point. The sniper team members will halt, ensuring they have good available concealment and know each other's location. The security element then proceed, leaving the sniper team in place. The sniper team will remain in position until the security element is well out of the area. The team will then organize itself as required by the mission and move on

to its tentative position. This type of separation also works well in MOUT situations.

f. When selecting routes, the sniper team must remember its vulnerabilities and limitations. The following guidelines should be used when selecting routes:

o Avoid known enemy positions and obstacles.

o Seek terrain that offers the best cover and concealment.

o Take advantage of difficult terrain (swamps, dense woods, and so forth).

o Avoid natural lines of drift.

o Do not use trails, roads, or footpaths.

o Avoid built-up or populated areas.

o Avoid areas of heavy enemy guerrilla activity.

o Avoid areas between opposing forces in contact with each other.

g. When the sniper team moves, it must always assume its area is under enemy observation. Because of this and the size of the team with the small amount of firepower it has, the team can use only one type of formation, the sniper movement formation. Characteristics of the formation are:

o The observer will be the point man, the sniper will follow.

o The observer's sector is 9 o'clock to 3 o'clock; the sniper's is 3 o'clock to 9 o'clock.

o Visual contact must be maintained, even when lying on the ground.

o An interval of no more than 20 meters is maintained.

o The sniper reacts to the point man's actions.

o The team leader designates the movement techniques and routes used.

o Danger areas are crossed by changing movement techniques.

h. A sniper team must never become decisively engaged with the enemy. The team must develop a set of drills that become a natural and immediate reaction should it make unexpected contact with the enemy. Examples of such actions are:

(1) <u>Visual contact</u>. If the sniper team sees the enemy and the enemy does not see the sniper team, it will freeze. If the team has time, it will do the following:

o Assume the best covered and concealed position.

o Remain in position until the enemy has passed.

o Will not initiate physical contact.

(2) <u>Ambush</u>. In an ambush, the sniper team's objective is to break contact immediately. One example of this is as follows:

(a) The observer delivers rapid fire on the enemy.

(b) The sniper throws smoke grenades between the observer and the enemy.

(c) The sniper then delivers well-aimed shots at the most threatening targets until smoke covers the area.

(d) The observer then throws fragmentation grenades and withdraws toward the sniper, ensuring not to mask the sniper's fire.

(e) The team moves to a location where the enemy cannot observe or place direct fire on them.

(f) If contact cannot be broken, the sniper calls for indirect fires/security element (if attached).

(g) If team members get separated, they should return to the next-to-last designated rally point.

(3) <u>Indirect fire</u>. When reacting to indirect fires, the team must move out of the area as quickly as possible. This sudden movement can result in the team's exact location and direction being pinpointed. Therefore, the team must not only react to indirect fire, but

also take actions to conceal its movement once it is out of the impact area.

(a) The team leader will move the team out of the impact area using the quickest route by giving the direction and distance (clock method).

(b) Both members will move out of the impact area the designated distance and direction.

(c) The team leader will then move the team farther away from the impact area by using the most direct concealed route. They will continue the mission using an alternate route.

(d) If team members get separated, they will return to the next-to-last designated rally point.

(4) <u>Air attack</u>.

o Team members assume the best available covered and concealed positions.

o Between passes of aircraft, team members will move to positions that offer better cover and concealment.

o Team will not engage the aircraft.

o Team members will remain in positions until attacking aircraft depart.

o If team members get separated, they will return to the next-to-last designated rally point.

i. To aid the sniper team members in navigation, they should memorize the route by studying maps, aerial photos, or sketches. Note distinctive features (hills, streams, roads) and their location in relation to the route. Plan an alternate route in case the primary route cannot be used. Plan an offset to circumvent known obstacles to movement. Use terrain countdown, which is memorizing terrain features from the start to the objective, to maintain the route. During the mission, the sniper team mentally counts each terrain feature as they cross it, thus ensuring they are maintaining the proper route.

j. The sniper team maintains orientation at all times. As they move, they observe the terrain carefully and mentally check off the distinctive features noted in the planning and study of the route. Many aids are available to ensure orientation.

o The location and direction of flow of principal streams.

o Hills, valleys, roads, and other peculiar terrain features.

o Railroad tracks, powerlines, and other man-made objects.

4-4. SELECTION AND OCCUPATION OF SNIPER POSITIONS

Selecting the location of a position is one of the most important tasks a sniper must accomplish during the mission planning phase of an operation. After selecting the location, the sniper must also determine how he will move into the area and locate and occupy the final position.

a. Selection of Positions. Upon receiving a mission, the sniper will locate the target area and then determine the best location for a tentative position by using one or more of the following sources of information: topographic maps, aerial photographs, visual reconnaissance before the mission, and information gained from units operating in the area.

(1) Once on the ground, the sniper will ensure the position provides an optimum balance between the following considerations:

o Maximum fields of fire and observation of the target area.

o Concealment from enemy observation.

o Covered routes into and out of the position.

o Located no closer than 300 meters from the target area.

o A natural or man-made obstacle between the position and the target area.

(2) A sniper must remember that a position that appears to be in an ideal location may also appear that way to the enemy. Therefore, the sniper will avoid choosing locations that are --

o On a point or crest of prominent terrain features.

o Close to isolated objects.

o At bends or ends of roads, trails, or streams.

o In populated areas, unless it is required.

(3) The sniper must use his imagination and ingenuity in choosing a good location for the given mission. He must choose a location that not only allows him to be effective, but it must also appear to the enemy to be the least likely place for a sniper's position. The following are examples of such positions:

o Under logs in a "dead-fall" area.

o Tunnels bored from one side of a knoll to the other.

o Swamps.

o Deep shadows.

o Inside rubble piles.

b. Occupation of Positions. During the mission planning phase, the sniper will also select an objective rally point, from which the sniper team will recon the tentative position to determine the exact location of its final position. The location of the ORP should provide cover and concealment from enemy fire and observation, be located as close to the selected area as possible, and have good routes into and out of the selected area.

(1) From the ORP, the team will move forward to a location that allows the team to view the tentative position area (Figure 4-8). One member will remain in this location and cover the other member while he recons the area to locate a final position. Once a suitable location has been found, the covering team member will move to the position. While conducting the recon or moving to the position, the team will --

o Move slowly and deliberately, using the sniper low crawl.

o Avoid unnecessary movement of trees, bushes, and grass.

o Avoid making any noises.

o Stay in the shadows if there are any.

o Stop, look, and listen every few feet.

(2) When the sniper team members arrive at the firing position, they will --

o Conduct a hasty and detailed search of the target area.

o Start construction of the firing position, if required.

o Organize equipment so that it is easily accessible.

o Establish a system of observing, eating, resting, and latrine calls.

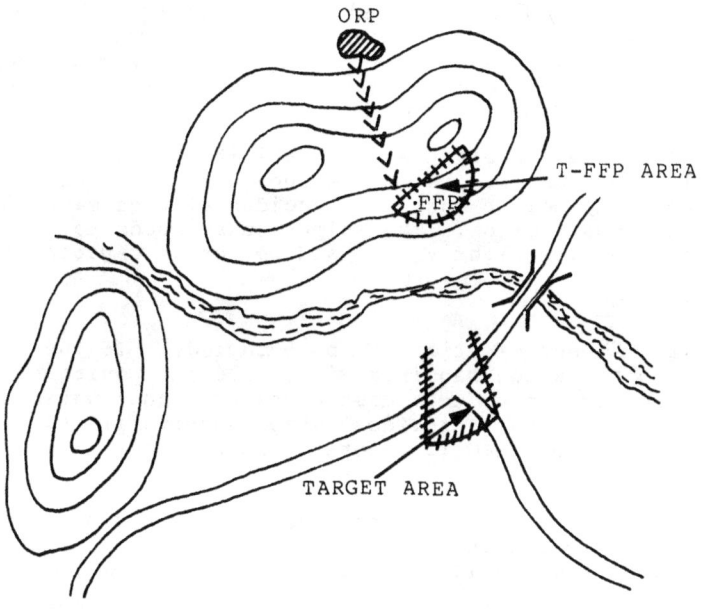

Figure 4-8. Tentative position areas.

4-5. CONSTRUCTION OF SNIPER POSITIONS

A sniper mission will always require the sniper to occupy some type of position. These positions can range from a hasty position, which a sniper may use for a few hours, to a more permanent position, which the team could remain in for a few days. When choosing and constructing positions, the sniper team must use their imagination and ingenuity to reduce the time and difficulty of position construction. The team should always plan to build their position during limited visibility.

   a. Sniper Position Considerations. Whether a sniper will be in a position for a few minutes or a few days, the basic considerations in choosing a type of position will be the same.

     (1) Location:

       (a) Type of terrain and soil. Digging and boring of tunnels can be very difficult in hard soil or in fine, loose sand. The sniper needs to take advantage of what the terrain offers (gullies, holes, hollow tree stumps, and so forth).

       (b) Enemy location and capabilities. Enemy patrols in the area may be close enough to the position to hear any noises that may accidentally be made during any construction. The sniper also needs to consider the enemy's night vision and detection capabilities.

     (2) Time:

       (a) Amount of time to be occupied. If the sniper team's mission requires it to be in position for a long time, the team must consider construction of a position that provides more survivability. This allows the team to operate more effectively for a longer time.

       (b) Time needed for construction. The time needed to build a position must be a consideration, especially during the mission planning phase.

     (3) Personnel and equipment:

       (a) Equipment needed for construction. The team needs to plan the use of any extra equipment needed for the construction (bow saws, picks, axes, and so forth).

(b) Personnel needed for construction. Coordination needs to take place if the position requires more personnel to build it or a security element to secure the area while the position is constructed.

b. Hasty Position. A hasty position is used when the sniper team will be in position for a short time, when they cannot construct a position due to the location of the enemy, or when the team must assume a position immediately. The hasty position is characterized by the following:

(1) Advantages:

(a) Requires no construction. The sniper team uses what is available for cover and concealment.

(b) Can be occupied in a short time. As soon as a suitable position is found, the team need only prepare loopholes, move small amounts of vegetation, or simply back a few feet away from the vegetation that is already there to conceal the weapon's muzzle blast.

(2) Disadvantages:

(a) No freedom of movement. This position does not allow the sniper team any free movement. Any movement that is not slow and deliberate may result in the team being compromised.

(b) Observation of large areas can be restricted. This type of position is normally used to observe a specific target point (intersection, passage, or crossing) rather than large open areas that require recons of the area to locate the best position for viewing a large area.

(c) No protection from direct or indirect fires. The team has no protection from indirect fires and only available cover for protection from direct fires.

(d) Must rely heavily on personal camouflage. The sniper's only protection against detection is his personal camouflage and his ability to use the available terrain.

(3) <u>Occupation time</u>. The team should not remain in this type of position longer than 8 hours. Remaining in this position longer than this will only result in loss of effectiveness due to muscle strain or cramps combined with eye fatigue because the position allows no freedom of movement.

c. Expedient Position. When a sniper team is required to remain in position for a longer time than the hasty position can provide, an expedient position (Figure 4-9) should be constructed. The expedient position lowers the sniper's silhouette as low to the ground as possible, but still allows him to fire and observe effectively. The expedient position is characterized by the following:

(1) <u>Advantages</u>:

(a) Requires little construction. This position is constructed by digging a hole out in the ground just large enough for the team and its equipment. Soil dug from this position can be placed in sandbags and used for making the firing tables.

(b) Conceals most of the body and equipment. The optics, rifles, and heads of the sniper team are the only items that are above ground level in this position.

(c) Provides some protection from direct fires. This position provides some protection from direct fires due to its lower silhouette.

(2) <u>Disadvantages</u>:

(a) Little freedom of movement. The team has more freedom of movement in this position than in the hasty position. However, snipers will need to remember that stretching a leg or reaching for a canteen will cause the exposed head to move unless controlled. The sniper can lower his head below ground level, but this should be done very slowly to ensure a target indicator is not produced.

(b) Little protection from indirect fires. This position will not protect the team from shrapnel and debris falling into the position.

(c) Head, weapons, and optics are exposed. The team must rely heavily on the camouflaging of these exposed items.

(3) <u>Construction time</u>: 1 to 3 hours (depending on the situation).

(4) <u>Occupation time</u>: 6 to 12 hours.

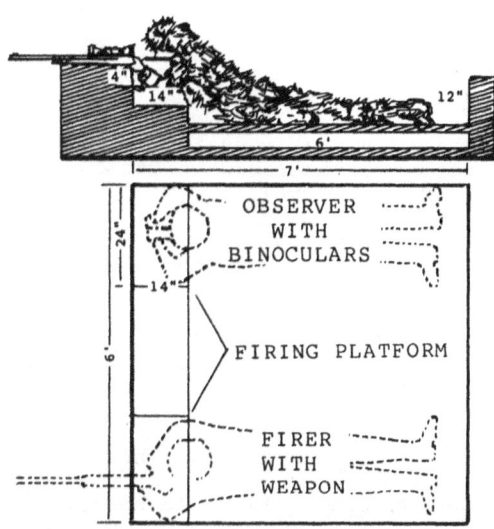

Figure 4-9. Expedient position.

d. Belly Hide. The belly hide (Figure 4-10) is similar to the expedient position, but it has overhead cover that not only protects the sniper from the effects of indirect fires, but also allows more freedom of movement. This position can be dug out under a tree, a rock, or any available object that will provide overhead protection and a concealed entrance and exit. The belly hide is characterized by the following:

(1) <u>Advantages</u>:

(a) Allows some freedom of movement. The darkened area inside this position allows snipers to move freely. Snipers must remember to cover the entrance/exit hole with a poncho or piece of canvas so outside light does not highlight the team inside the position.

(b) Conceals all but the rifle barrel. All equipment is inside the position except the rifle barrels, but the barrels could be inside, depending on the room available to construct the position.

(c) Provides protection from direct and indirect fires. The team should try and choose a position that has an object that will provide good overhead protection (rock, tracked vehicle, rubble pile, and so forth), or prepare it in the same manner as overhead cover for other infantry positions.

(2) <u>Disadvantages</u>:

(a) Requires extra construction time. If the team has to construct overhead cover for the position, it will require more time.

(b) Requires extra materials and tools. Construction of overhead cover will require saws or axes, waterproof material, and so forth.

(c) Cramped space. The sniper team will have to lay in the belly hide without a lot of variation in body position due to limited space and design of the position.

(3) <u>Construction time</u>: 4 to 6 hours.

(4) <u>Occupation time</u>: 12 to 48 hours.

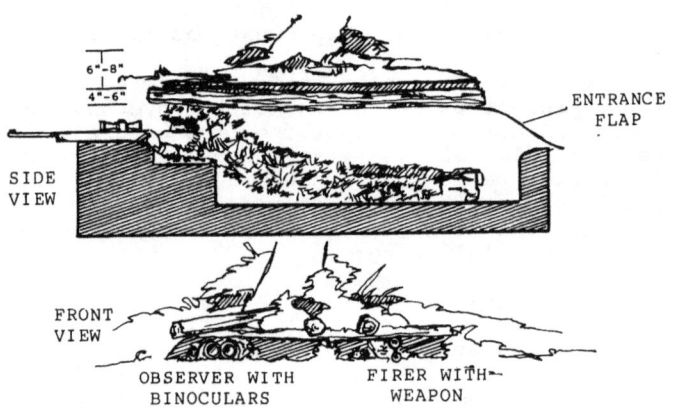

Figure 4-10. Belly hide position.

e. Semipermanent Hide. The semipermanent hide (Figure 4-11) is used mostly in a defensive or outpost situation. This position requires additional equipment and personnel to construct, but will allow sniper teams to remain there for extended periods or be relieved in place by other sniper teams. Like the belly hide, this position can be constructed by tunneling through a knoll or under natural objects already in place. The semipermanent hide is characterized by the following:

(1) Advantages:

(a) Total freedom of movement inside the position. The team members are able to move about freely. They can stand, sit, or even lie down.

(b) Protection from direct and indirect fires. The sniper team should look for the same items as mentioned in the belly hide.

(c) Completely concealed. Loopholes are the only part of the position that can be detected. Loopholes allow for the smallest exposure possible. Yet, they still allow the sniper and observer to view the target area. These loopholes should have a large diameter (10 to 14 inches) in the interior of the position and taper down to a smaller diameter (4 to 8 inches) on the outside of the position. A position may have more than two loopholes if needed to cover large areas/to see to the rear of the position. The entrance/exit to the position must be covered to prevent light from entering and highlighting the loopholes. Loopholes that are not in use should be covered from the inside with a piece of canvas or suitable material.

(d) Can be maintained for extended periods. This position allows the team to operate effectively for a longer period.

(2) Disadvantages:

(a) Requires extra personnel and tools to construct. A position like this will require extensive work and more tools. Very seldom can a position like this be constructed in the close vicinity of the enemy, but it should be constructed during the hours of darkness and be completed before dawn.

(b) Using a position for several days or having teams relieve each other in a position will always increase the risk of the position being detected. Snipers should never continue to fire from the same position.

(3) <u>Construction time</u>: 4 to 6 hours (4 personnel).

(4) <u>Occupation time</u>: 48 hours plus (relieved by other teams).

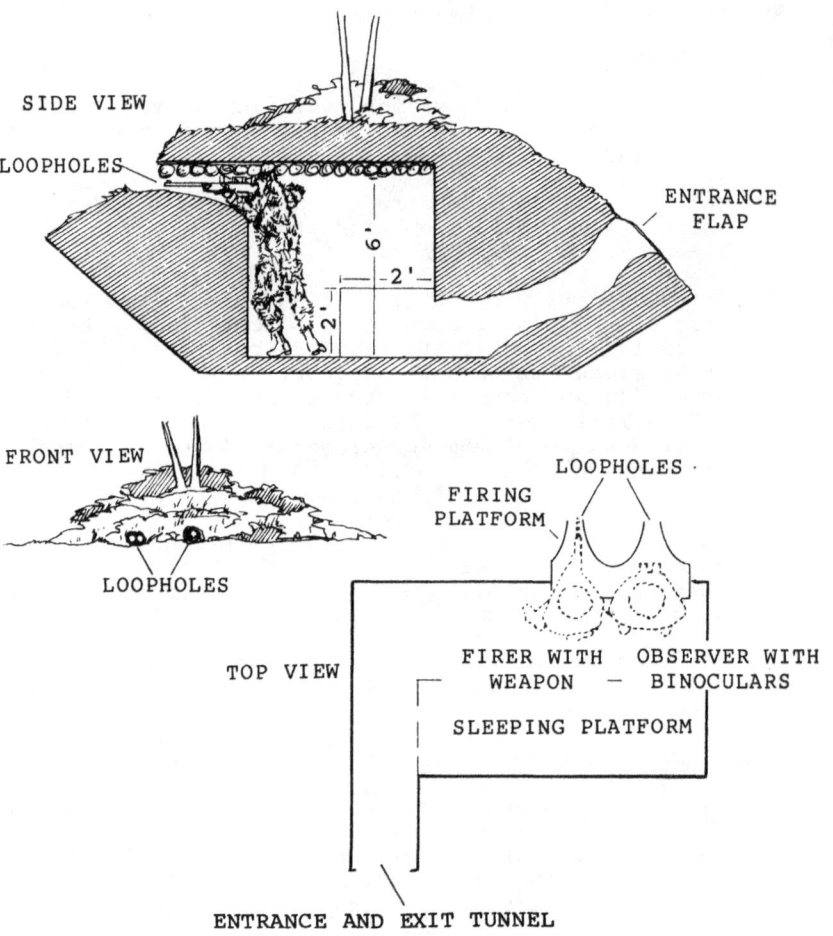

Figure 4-11. Semipermanent hide position.

f. **Routines in Sniper Positions.** Although the construction of positions may differ, the routines while in position are the same. The sniper and the observer should have a good firing platform. This gives the sniper a stable platform for the sniper weapon and the observer a platform for the optics. When rotating observation duties, the sniper weapon should remain in place, and the optics are handed from one member to the other. Data books, observation logs, range cards, and the radio should be placed in between the team where both members have easy access to them. A system of resting, eating, and latrine calls must be arranged between the team. All latrine calls should be done during the hours of darkness, if possible. A hole should be dug to conceal any traces of latrine calls.

g. **Positions in Urban Terrain.** Positions in urban terrain are quite different than positions in the field. The sniper team will normally have several places to choose from. These can range from inside attics to street level positions in basements. This type of terrain is ideal for a sniper, and a sniper team can literally stop an enemy's advance through its area of responsibility. But, one important fact for the sniper to remember is that in this type of terrain, the enemy will use every asset he has to detect and eliminate a sniper. When constructing an urban position, the sniper must --

(1) Always be aware of the outside appearance of the structure he is in. Shooting through loopholes in barricaded windows is preferred; but, make sure all the other windows are also barricaded. Placing loopholes in these other windows also provides more than one position to fire from. When making loopholes, make them different shapes (not perfect squares or circles). Dummy loopholes will also confuse the enemy. Positions in attics are also very effective. Remove shingles and cut out loopholes in the roof; however, make sure there are other shingles missing from the roof so the firing position loophole is not obvious.

(2) Not locate his position against contrasting background or in prominent buildings that automatically draw attention. He must stay in the shadows while moving, observing, and engaging targets.

(3) Never fire close to a loophole. Always back away from the hole as far as possible to hide the muzzle flash and scatter the sound of the weapon when it fires.

Some positions can be located in a different room than the one the loophole is in by making a hole through a wall connecting the two and firing from inside the far room. Do not fire continually from one position. (This is why more than one position should be constructed if time and situation permit.) When constructing other positions, make sure the target area can be observed. Sniper positions should never be used by any personnel other than snipers.

4-6. OBSERVATION AND TARGET SELECTION

Snipers must be able to detect, identify, describe, and plot the location of enemy personnel and equipment on the battlefield. To do this, the sniper must develop good techniques of observation. Observation is a planned, systematic process of viewing an area for indications of enemy presence or activity. It is an ongoing task that must be conducted every minute the sniper team is in position, because the time the area is not being observed could be the only time the enemy presents himself or a target indicator. While observing a target area, the sniper will alternately conduct two types of searches, a hasty search and a detailed search.

   a. A hasty search is the first phase of observing a target area and is conducted by the observer immediately after the team occupies the firing position. A hasty search is quick glances with binoculars at specific points, terrain features, or other areas that could conceal the enemy. The area closest to the sniper position should be viewed first, because it could pose the most immediate threat. The observer will then search farther out until the entire target area has been searched. Only when the observer sees or suspects he sees a target will he use the M49 telescope which gives him a more detailed view of the target area. The telescope should not be used to search the area because its narrow field of view would take him much longer to cover an area; plus, its stronger magnification will cause eye fatigue sooner than the binoculars will.

   b. After a hasty search has been completed, the observer will then conduct a detailed search of the area. A detailed search is a closer, more thorough search of the target area, using 180-degree arcs or sweeps, 50 meters in depth, and overlapping each previous sweep at least 10 meters to ensure the entire area has been observed (Figure 4-12). Like the hasty search, the observer will begin by searching the area closest to the sniper team position.

4-28

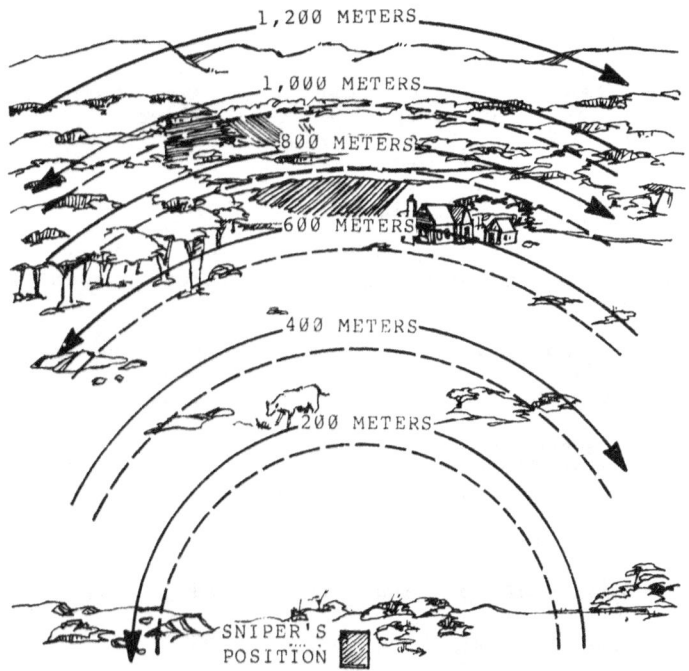

Figure 4-12. Detailed search.

c. This cycle of a hasty search followed by a detailed search should initially be repeated three or four times. This allows the sniper to become accustomed to the area; plus, he will tend to look closer at various points with each consecutive pass over the area. After the initial searches, the observer should view the area using a combination of both hasty and detailed searches. While the observer is conducting the initial searches of the area, the sniper should record prominent features, reference points, and distances on a range card. The team members should alternate the task of observing the area about every 30 minutes.

d. The team will record any targets observed in the area for future reference. Recording the types and location of targets observed in the area will aid the team in determining targets to be engaged. The sniper must be able to select the key targets that will do the greatest harm to the enemy in the given situation. Snipers must also consider the use of indirect fire on targets. Some targets, due to their size or location, may be better engaged with indirect fire. The sniper may use indirect fires to disguise fire from his position or to help cover his withdrawal after engaging a target. When selecting key targets, the sniper must consider the following factors:

(1) Threat to the sniper. The sniper must consider the danger the target presents. This can be an immediate threat, such as an enemy element walking upon the sniper's position; or a future threat, such as enemy snipers or dog tracking teams.

(2) Probability of first-round hit. The sniper must determine the chances of hitting the target with the first shot by considering the following:

o Distance to the target.

o Direction and velocity of the wind.

o Visibility of the target area.

o Amount of the target that is exposed.

o Amount of time the target is exposed.

o Speed and direction of target movement.

(3) Certainty of target's identity. The sniper must be reasonably certain that the target he is considering is indeed the key target that he wants.

(4) Target's impact on the enemy. The sniper must consider what impact the elimination of the target will have on the enemy's fighting ability. The sniper must determine that the target is the one available target that will cause the greatest harm to the enemy.

(5) Enemy's reaction to sniper fire. The sniper must consider what the enemy will do once the shot has been fired. The sniper must be prepared for such actions as immediate suppression by indirect fires and enemy sweeps of the area.

(6) Effect on the overall mission. The sniper must consider how the engagement will affect his overall mission. The mission may be one of intelligence gathering for a certain period, and firing will not only alert

the enemy to a sniper's presence but may also terminate the mission if the sniper has to move from the position as a result of the engagement.

e. Key personnel targets can be identified by actions or mannerisms, positions within formations, rank or insignias, and or equipment being worn or carried. Key targets can also include weapon systems and equipment. Examples of key targets are:

   (1) Snipers. Snipers are the number one target of a sniper team. The enemy sniper not only poses a threat to friendly forces, he is the natural enemy of "you" the sniper, and he can stalk you on your own terms. The fleeting nature of a sniper is reason enough to engage him because you may never see him again.

   (2) Dog tracking teams. Dog tracking teams pose a great threat to sniper teams and other special teams that may be working in the area. It is very hard to fool a trained dog's nose; therefore, the team must be stopped. When engaging a dog tracking team, the sniper should engage the dog's handler first. This confuses the dog and he may not be controllable by the other members of the team.

   (3) Scouts. Scouts are keen observers and provide valuable information about friendly units. This plus their ability to control indirect fires make them very dangerous on the battlefield. They must be eliminated.

   (4) Officers (military and political). Officers are another key target of the sniper. Losing key officers in some forces is such a blow to their operating capability that they may not be able to make a coordinated effort for hours.

   (5) NCOs. Losing NCOs not only affects the operation of a unit, but also affects the morale of the lower ranking personnel.

   (6) Vehicle commanders and drivers. Many vehicles are rendered useless without a commander or driver.

   (7) Communications personnel. In some forces, only highly trained personnel know how to operate various types of radios. Eliminating these personnel can be a serious blow to the enemy's communication network.

(8) **Weapon crews**. Eliminating weapon crews reduces the amount of fire on friendly troops.

(9) **Optics on vehicles**. Personnel who are in closed vehicles are limited to viewing through optics. The sniper can blind a vehicle by damaging these optic systems.

(10) **Communication and radar equipment**. The right shot in the right place can completely ruin a tactically valuable radar or communication system. Plus, only highly trained personnel may attempt to repair these systems in place. Eliminating these personnel may cause a serious blow to the enemy's field repair capabilities.

(11) **Weapon systems**. Many high-tech weapons, especially computer-guided type systems, can be rendered useless by one well-placed round in the guidance controller part of the system.

4-7. INFORMATION RECORDS

The secondary mission of the sniper is the collection and reporting of information. To accomplish this, the sniper not only needs to be a keen observer, but also must be able to accurately relay the information he has observed. To record this information, the sniper uses range cards, military sketches, and the observation logbook.

  a. Range Cards. The range card represents the target area, drawn as seen from above with annotations indicating distances throughout the target area (Figure 4-13). See Appendix G for the blank reproducible form of the Sniper's Range Card. The range card gives the sniper a quick range reference and a means to record target locations since it has preprinted range rings on it. These cards can be broken into sectors by using dashed lines (Figure 4-14). This gives the team members a quick reference when locating targets. Example: "The intersection in sector A." A field expedient range card can be prepared on any paper the team has available. The sniper position and distances to prominent objects and terrain features will be drawn on the card. There is not a set maximum range on either range card because the sniper may also label any indirect fire targets on his range card. Information contained on both range cards includes:

o Name and method of obtaining range.

o Left and right limits of engageable area.

o Major terrain features, roads, and structures.

o Ranges, elevation, and windage needed at various distances.

o Distances throughout the area.

o Temperature and wind. (Cross out previous entry whenever temperature, wind direction, or wind velocity changes.)

o TRPs (azimuth, distance, and description).

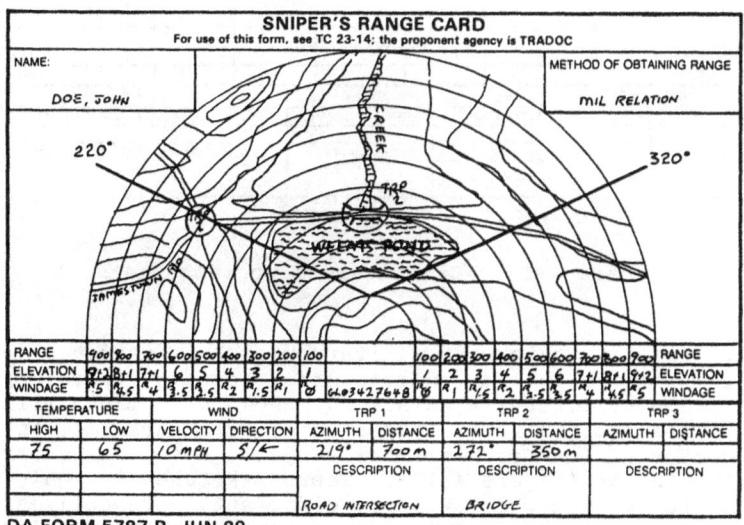

Figure 4-13. Prepared range card.

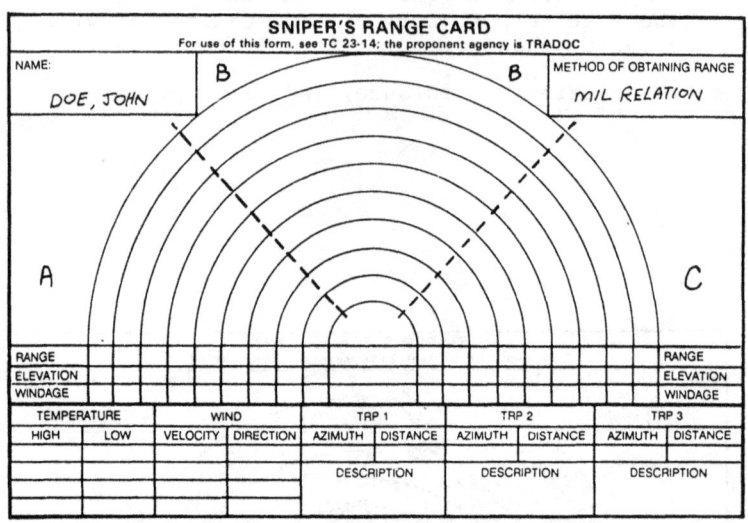

Figure 4-14. Dashed lines on range card.

b. **Military Sketch.** A military sketch is used to record information about a general area, terrain features, or man-made structures that are not shown on a map. Military sketches provide intelligence sections a detailed, on the ground view of an area or object that is otherwise unobtainable (Figure 4-15). These sketches not only let the viewer see the area in different perspectives but provide detail such as type of fences, number of telephone wires, present depth of streams, and so forth. There are two types of military sketches as stated in FM 21-26: road/area sketches and field sketches.

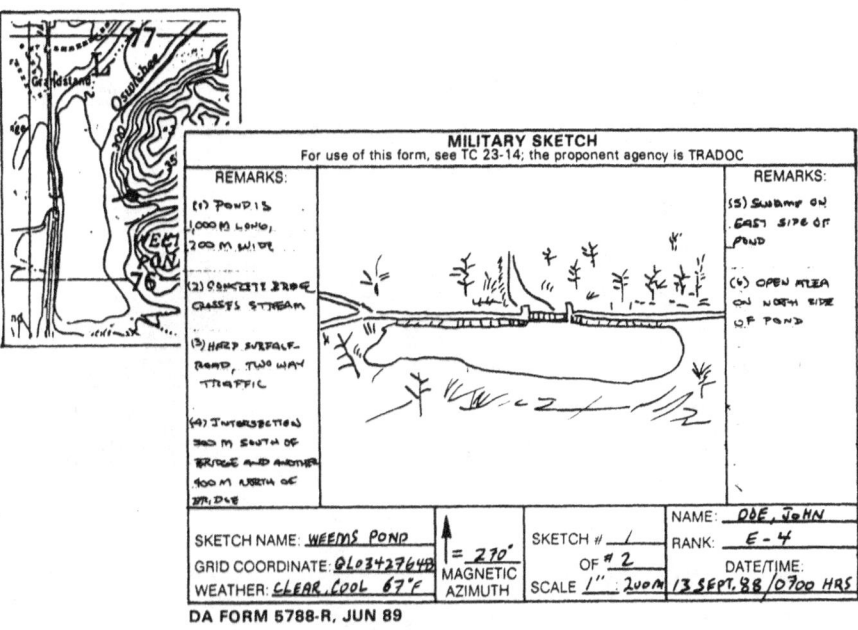

Figure 4-15. Military sketch.

(1) <u>Road/Area</u>. A road/area sketch (Figure 4-16) is a panoramic representation of an area or object drawn to scale as seen from the sniper's perspective. It depicts detailed information about a specific area or a man-made structure. The sketch includes:

o   Grid coordinates of sniper's position.

o   Magnetic azimuth through the center of sketch.

o   Sketch name and number.

o   Scale of sketch.

o   Remarks section.

o   Name and rank.

o   Date and time.

o   Weather.

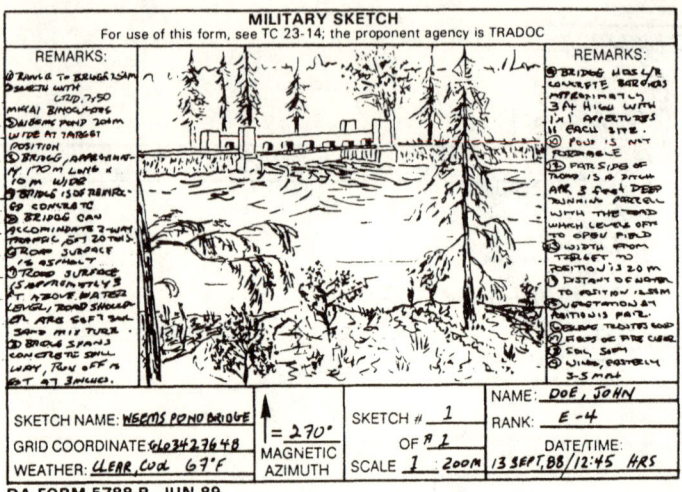

Figure 4-16. Road/area sketch.

(2) <u>Field sketch</u>. A field sketch (Figure 4-17) is a topographic representation of an area drawn to scale as seen from above. It gives the sniper a method of describing large areas while showing reliable distance and azimuths between major features. This type of sketch is useful in describing road systems, flow of streams/rivers, or locations of natural and man-made obstacles. The field sketch can also be used as an overlay on the range card. Information contained in a field sketch includes:

o   Grid coordinates of sniper's position.

o   Left and right limits with azimuths.

o   Rear reference with azimuth and distance.

o   Target reference points.

o   Sketch name and number.

o   Name and rank.

o   Date and time.

o   Weather and visibility.

4-36

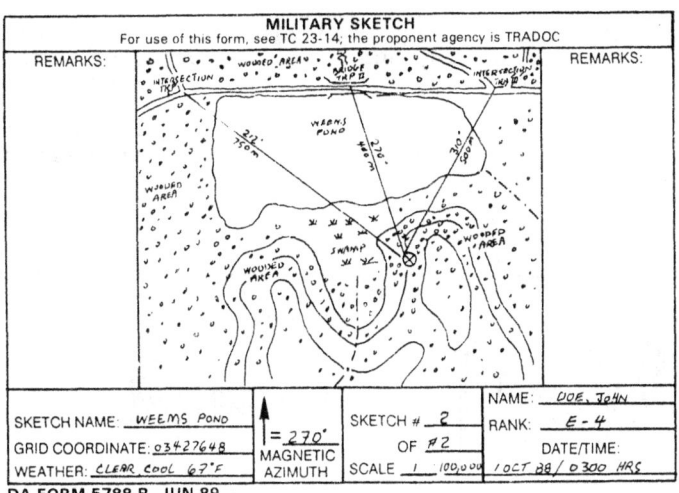

Figure 4-17. Field sketch.

(3) <u>Drawing sketches</u>. As with all drawings, artistic skill is an asset, but satisfactory sketches can be produced by anyone with enough practice. The following are guidelines to be used when drawing sketches:

(a) Work from the whole to the part. The sniper must first determine the boundaries of the sketch. He should then sketch the larger objects such as hills, mountains, or outlines of large buildings. Only after drawing the large objects in the sketch should the sniper start drawing the smaller details.

(b) Use common shapes to show common objects. The sniper does not need to sketch each individual tree, hedgerow, or woodline exactly. He should use common shapes to show these types of objects. Do not concentrate on the fine details unless they are of tactical importance.

(c) Draw in perspective; use vanishing points. All sketches should be drawn in perspective. To do this, the sniper must recognize the vanishing points of the

4-37

area to be sketched. Parallel lines on the ground
that are horizontal vanish at a point on the horizon
(Figure 4-18). Parallel lines on the ground that
slope downward, away from the observer vanish at a
point below the horizon. Parallel lines on the ground
that slope upward, away from the observer vanish at a
point above the horizon. Parallel lines that recede
to the right vanish on the right, and those that
recede to the left vanish on the left. An example of
the steps in preparing a road/area sketch are shown in
Figure 4-19.

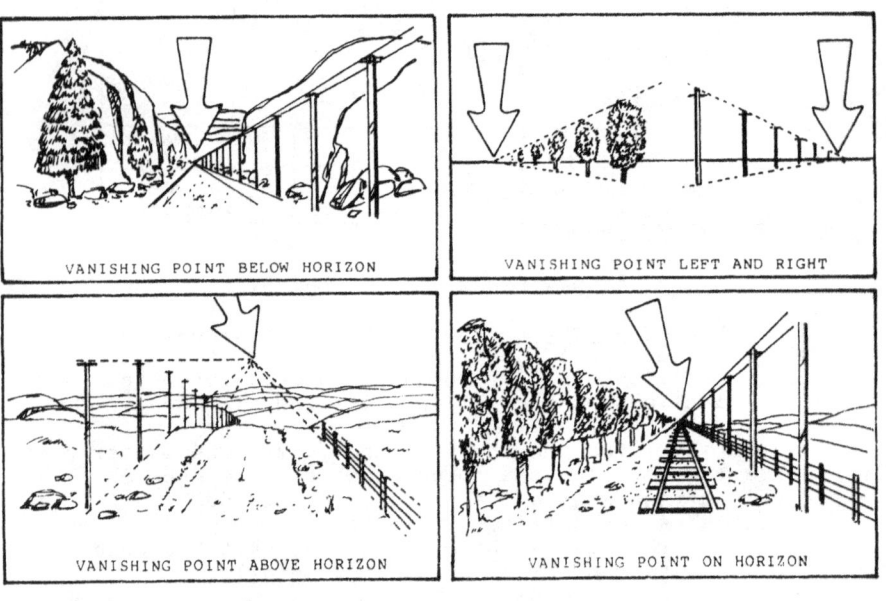

Figure 4-18. Vanishing points.

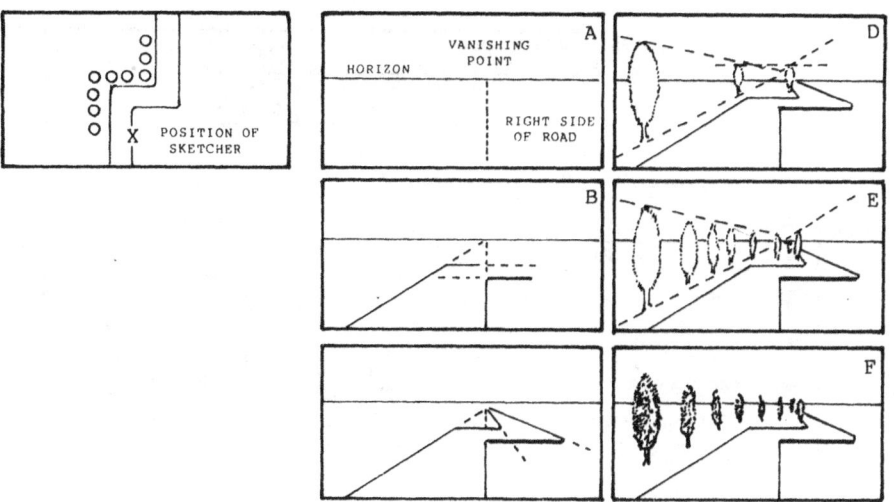

Figure 4-19. Preparing a road/area sketch.

c. Logbook. The observation logbook (Figure 4-20) is a written, chronological record of all activities and events that take place in a sniper team's area. It is used in conjunction with military sketches and range cards; this combination not only gives commanders and intelligence personnel information about the appearance of the area, but it also provides an accurate record of the activity in the area. The data in the observation logbook includes:

o Grid coordinates of sniper's position.

o Observer's name.

o Date and time of observation and visibility.

o Sheet number and number of total sheets.

o Series number, time, and grid coordinates of each event.

o The event that has taken place.

o Action taken.

SNIPER'S OBSERVATION LOG					
For use of this form, see TC 23-14; the proponent agency is TRADOC				SHEET 5 OF 5 SHEETS	
ORIGINATOR: DOE, JOHN			DATE/TIME: 1 OCT 88	LOCATION: GL0342764 8	
SERIAL	TIME	GRID COORDINATE	EVENT	ACTIONS OR REMARKS	
1	0300	GL03427698	OCCUPIED POSITION	OBSERVATION	
2	0340	SAME	PFC GOD RESTED	NONE	
3	0430	SAME	PFC GOD ASSUMED OBSERVATION DUTIES	I RESTED	
4	0520	SAME	BOTH OF US AWAKE	NONE	
5	0630	SAME	PREPARED RANGE CARD + TOPOGRAPHIC SKETCH	LIGHT ENOUGH TO SEE	
6	0655	GL03117631	BMND CROSSED BRIDGE GOING SOUTH	OBSERVED	
7	0700	GL03427648	PREPARED SKETCH OF BRIDGE GL03117631	COMPLETE	
8	0900	GL03427648	MISSION COMPLETED/ RETURN TO CP	END OF MISSION	

DA FORM 5786-R, JUN 89

Figure 4-20. Prepared sheet from observation logbook.

4-8. RANGE ESTIMATION

A sniper is required to determine distance accurately in order to properly adjust elevation on the sniper weapon system and to prepare topographical sketches or range cards. Because of this, the sniper has to be proficient in the various range estimation techniques.

a. Paper Strip Method. The paper strip method (Figure 4-21) is useful when determining longer distances (1,000 meters plus). When using this method, the sniper will place the edge of a strip of paper on the map, ensuring it is long enough to reach between the two points, and pencil in a tick mark on the paper at the sniper's position and another at the distant location. The sniper will then place the paper on the map's bar scale, located at the bottom center of the map, and align the left tick mark with the 0 on the scale. Then he reads to the right where the second mark is and notes the corresponding distance represented between the two marks.

4-40

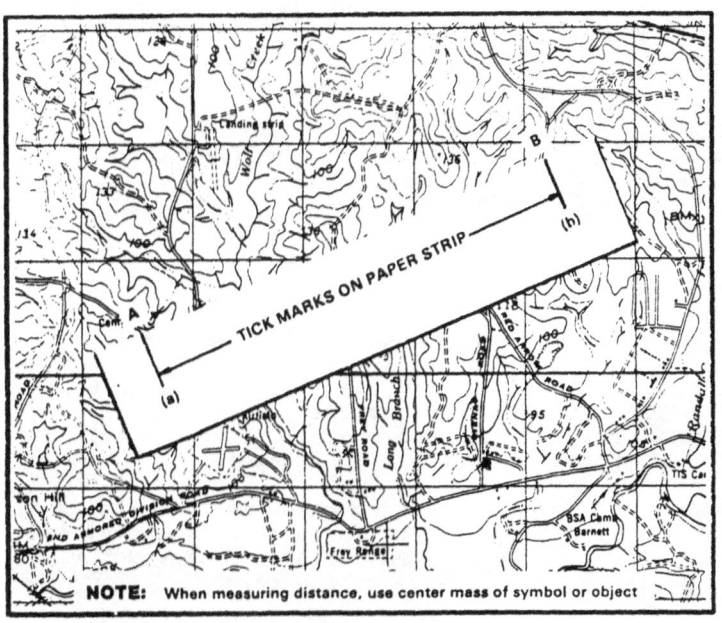

Figure 4-21. Paper strip method.

b. 100-Meter-Unit-of-Measure Method. To use this method (Figure 4-22), the sniper must be able to visualize a distance of 100 meters on the ground. For ranges up to 500 meters, he determines the number of 100-meter increments between the two objects he wishes to measure. Beyond 500 meters, the sniper must select a point halfway to the object and determine the number of 100-meter increments to the halfway point, then double it to find the range of the object.

4-41

Figure 4-22. 100-meter-unit-of-measure method.

c. Appearance-of-Object Method. This method is a means of determining range by the size and other characteristic details of the object. To use the appearance-of-object method with any degree of accuracy, the sniper must be thoroughly familiar with the characteristic details of the objects as they appear at various ranges.

d. Bracketing Method. Using this method, the sniper assumes that the target is no more than X meters, but no less than Y meters away. An average of X and Y will be the estimate of the distance to the target.

e. ART I or II scopes. The scopes on the M21 sniper weapon system can also be used for rough range estimation. Once the sniper is familiar with his M21 and is accustomed to ranging out on targets, he makes a mental note of where the power adjust ring is set at various distances.

f. Range Card Method. The sniper can also use a range card to quickly determine ranges throughout the target area. Once a target is seen, the sniper determines where it is located on the card and then reads the proper range to the target.

g. Mil-Relation Formula. The mil-relation formula is the preferred method of range estimation. This method is conducted by using a mil-scale reticle found in the M19 binoculars (Figure 4-23) or in the M3A sniperscope (Figure 4-24). To use this method, the sniper must know the target's size in inches or meters. Once the target size is known, the sniper then compares the target's size to the mil-scale reticle and uses the following formula:

$$\frac{\text{Size of target in meters} \times 1,000}{\text{Size of object in mils}} = \text{Range to target in meters}$$

(To convert inches to meters, multiply the number of inches by .0254.)

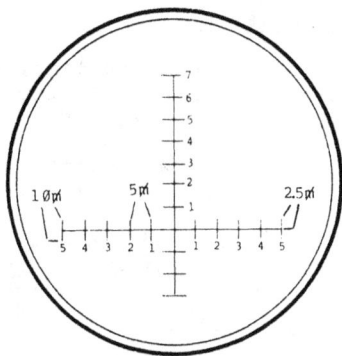

Figure 4-23. M19 binocular reticle.

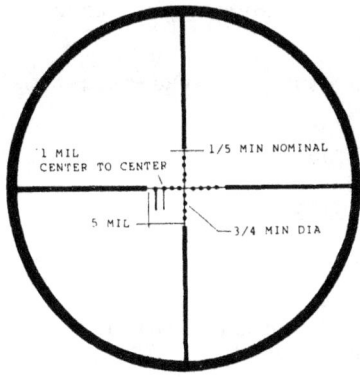

Figure 4-24. M3A sniperscope reticle.

h. **Combination Method.** In a combat environment, perfect conditions rarely exist, so just one method of range estimation may not be sufficient for your particular mission. Terrain with a lot of dead space limits the accuracy of the 100-meter method. Poor visibility limits the use of the appearance-of-object method. By using the combination of the two methods or more to determine an unknown range, an experienced sniper should arrive at an estimated range close to the true range.

i. **Factors Affecting Range Estimation.** There are three major factors that affect range estimation.

   (1) <u>Nature of the target</u>.

      (a) An object of regular outline, such as a house, will appear closer than one of irregular outline, such as a clump of trees.

      (b) A target that contrasts with its background will appear to be closer than it actually is.

      (c) A partially exposed target will appear more distant than it actually is.

   (2) <u>Nature of the terrain</u>.

      (a) As the observer's eye follows the contour of the terrain, he will tend to overestimate distant targets.

      (b) Observing over smooth terrain, such as sand, water, or snow, causes you to underestimate distant targets.

      (c) Looking downhill, the target will appear farther away.

      (d) Looking uphill, the target will appear closer.

   (3) <u>Light conditions</u>.

      (a) The more clearly a target can be seen, the closer it appears.

      (b) When the sun is behind the viewer, the target appears to be closer.

      (c) When the sun is behind the target, the target is more difficult to see and appears to be farther away.

TC 23-14

CHAPTER 5
EMPLOYMENT

The sniper delivers long-range precision fire on key targets and collects and reports battlefield information. A sniper team properly employed can disrupt enemy movement, observation, and infiltration; plus, influence the enemy's decisions and actions. It can instill fear and cause confusion, thus lowering the enemy's morale. When employing snipers, employment officers and sniper team leaders should use common sense and imagination. They must remember that a sniper mission cannot be tied to a rigid time schedule, nor can a sniper team be effective in positions that will receive fires from the enemy. The team must be positioned far enough away to avoid these fires, but close enough to deliver precision fires at the most threatening targets.

5-1. SNIPER TEAMS

Snipers are trained as teams and should always be employed as teams. Team members assist each other during long periods of observation and with range estimations, adjustments of rounds, and security.

   a. Sniper teams should be employed directly under the commander or the sniper employment officer. Snipers must know the commander's intent, scheme of maneuver, and fire support plan. Teams must be able to choose their own positions once they are on the ground to ensure the position will allow clear fields of fire and observation. Snipers cannot be used effectively in areas that do not allow maximum fields of fire and observation. The number of sniper teams participating in an operation will depend on their availability, the expected duration of the mission, and the enemy strength.

   b. Snipers cannot afford to be detected, so their movement is slower than other units. Therefore, they must be employed ahead of any anticipated movement. Sniper teams should move with a security element (squad or platoon). This allows the sniper team to reach positions in less time; plus, the security element provides a reaction force if the team is compromised.

   c. Snipers are affected by the following METT-T factors:

   o Mission - Type of mission team is required to accomplish.

5-1

o Enemy - Characteristics and capabilities of the enemy forces.

o Terrain - Type and condition of terrain, trails, roads, or water crossings.

o Troops available - Any other teams or security elements.

o Time - Time allowed/needed for the mission.

5-2. OFFENSIVE EMPLOYMENT

Offensive operations are designed to carry the fight to the enemy to destroy his capability/will to fight. The sniper can play a major role in offensive operations by killing enemy targets that threaten the success of the mission.

a. During offensive operations snipers --

o Engage enemy snipers.

o Overwatch movement and fire on targets threatening movement.

o Place precision fire on enemy crew-served weapons.

o Place precision fire on enemy leaders.

o Place precision fire into exposed apertures of bunkers.

o Place fire on bypassed forces.

o Fire at targets threatening a counterattack or fire at fleeing personnel.

o Provide flank protective fire at targets threatening an exposed flank.

o Dominate key terrain.

b. When employing snipers in a movement to contact, they can move with the lead element or be employed 24 to 48 hours before the units movement --

o To select positions.

o To gather information on the enemy.

o To dominate key terrain to prevent enemy surprise attacks.

c. The rapid movement of a mounted attack limits the sniper's role. However, when the unit dismounts, the sniper can be employed to support the assault.

d. During a raid, sniper teams can be employed with either the security element or the support element --

o To cover avenues of approach and escape into and out of the objective.

o To cover routes of friendly withdrawal to the rally point.

o To provide long-range fires on the objective.

5-3. DEFENSIVE EMPLOYMENT

In the defense, snipers will systematically shoot the specific targets that will cause the greatest hindrance to the enemy's advance.

a. During defensive operations, snipers --

o Cover obstacles, minefields, and demolitions.

o Kill enemy reconnaissance elements.

o Engage enemy armored-vehicle commanders exposed in turrets, antitank teams, and OPs.

o Shoot vehicle optics to slow movement.

o Place fire on enemy crew-served weapons.

o Place fire on enemy follow-up units.

b. In a reverse slope defense, snipers can provide effective long-range fire from positions forward of the topographical crest.

5-4. RETROGRADE EMPLOYMENT

During the delay and withdrawal, snipers can cause the enemy to deploy prematurely by inflicting casualties with accurate, long-range fire.

5-5. MOUT EMPLOYMENT

Urban environments present a variety of difficulties not encountered on the conventional battlefield: the terrain is mostly artificial; and streets and highways clearly define observation areas and field of fire. The observation and target detection capabilities combined with precision fire gives the sniper a major role in urban operations.

    a. Building interiors and underground passages are the best routes of movement since movement through streets is easily detected.

    b. Snipers are preferably positioned in buildings of masonry construction. These offer the best protection, long-range fields of fire, and all-round observation. Fields of fire are obvious to the enemy, and the sniper should muffle the sound and conceal the muzzle flash.

    o Align shots through openings in adjacent buildings.

    o Fire through funnel-shaped hole in a wall with the large end of the funnel at the room's interior.

    c. Building-to-building and street-to-street fighting shortens expected engagement ranges for snipers.

    d. Snipers in MOUT offense and defense will engage similar targets as discussed in paragraphs 5-2 and 5-3.

5-6. COUNTERSNIPER OPERATIONS

Countersniper operations are designed to eliminate enemy snipers. These operations must be thoroughly planned by the sniper teams involved. This is a battle between two highly trained elements, each knowing the capabilities and limitations of the other.

    a. **Determining a Sniper Threat.** The first task a sniper must accomplish in a countersniper operation is to determine if there is indeed a sniper threat. In doing this, the sniper identifies the following information gained from the unit operating in the area.

        (1) Unit personnel have seen enemy soldiers wearing special camouflage uniforms.

        (2) Unit personnel have seen enemy soldiers carrying weapons that have long barrels, mounted scopes, bolt

action receivers, or are carried in weapon cases or drag bags.

(3) The unit has had single shot reduction of key personnel (commanders, platoon leaders, senior NCOs, or weapon crews).

(4) There were marked reductions in enemy patrolling activities during the times of single-shot reductions.

(5) Unit personnel have detected reflections of light off optical lenses.

(6) Intelligence or recon patrols have reported small groups of enemy personnel (1 to 3 men), through visual sightings or tracking.

(7) There has been a finding of single expended casings such as the 7.62 x 54R-mm (same round used in some light machine guns).

b. Planning for a Countersniper Operation. Once the sniper has determined that an enemy sniper is operating in the area, he must determine the best method to use in eliminating the enemy sniper. To do this he will --

(1) Gather information. He must learn the times of the day reductions occurred. Then he finds out the locations that enemy sniper fire has been encountered and of any enemy sniper sightings. He gathers any material evidence of enemy snipers (casings, equipment, and so forth).

(2) Determine any patterns. The sniper will evaluate the information gathered to detect any patterns or routines the enemy sniper has established. The sniper should conduct map recons, study aerial photos, or conduct a ground recon to determine any patterns in travel. The sniper must place himself in the enemy's shoes and ask, "How would I accomplish the mission if I were him?"

(3) Plan actions. Once a pattern or routine is detected, the sniper will determine the best location and time to engage the enemy sniper. The sniper should also request the following actions:

(a) Coordination of routes and fires with the unit in the area.

(b) Additional pre-plotted targets (fire support).

(c) Infantry support to canalize or ambush the sniper.

(d) Additional sniper teams for mutual supporting fire.

(e) Someone to bait likely engagement areas to deceive the enemy sniper into committing himself by firing. An example of this would be to use a ghillie suit to create a dummy sniper. Place it in an open area to entice the enemy into engaging it; this will give the sniper a chance to detect the enemy's location.

(f) All elements be in place no later than 12 hours before the expected engagement time.

During a countersniper operation, the sniper must ignore the larger battle going on around him. He must concentrate on his one objective -- the enemy sniper.

c. Passive Countersniper Actions. When an enemy sniper is operating in a unit's area, the unit should employ passive measures to defend against sniper fire. Examples of these measures are:

(1) Do not stick to consistent routines, such as chow times, ammo resupply times, assembly area procedures, or any day-to-day activities.

(2) Conduct all meetings, briefings, or any gathering of people under cover or during limited visibility.

(3) Cover or conceal all equipment.

(4) Remove rank from helmets and collars. Do not salute officers. Leaders should not present authoritative mannerisms.

(5) Increase the unit's observation capabilities, such as OPs.

(6) Brief patrols to look for single expended rounds, different camouflage materials, and so forth.

(7) While performing the above actions, do not make it apparent that you are aware of a sniper's presence.

(8) Do not overlook women. An estimated 50 percent of snipers in many third world countries are women. Patrols and OPs should not be misled if they see a woman with a scoped rifle. She is a deadly opponent.

TC 23-14

APPENDIX A
FIELD TRAINING EXERCISES

Sniper training exercises provide a sniper with practical experience in detecting and engaging realistic targets under field conditions on ranges comparable to a battlefield. This training also provides a sniper with a means to practice the various sniper training fundamentals he has been taught previously, often collectively. These exercises may or may not be graded; however, competition is a proven method to obtain desired results. At the end of the exercises, the instructor will critique the sniper on his performance to include observation, range estimation, concealment, concealed movement, and rifle firing. These exercises include:

o Zeroing and practice fire.

o Field firing.

o Observation.

o Range estimation.

o Concealment and concealed movement.

o Land navigation.

o Record exercises.

o KIM games.

A-1. TRAINING NOTES

During all field training exercises, the individual sniper should be equipped as indicated in Chapter 2. Team equipment should be available as needed.

a. A standard known-distance range, graduated in 100-meter increments from 100 to 1,000 meters, is required for the zeroing exercises. The target detection range facilities and procedures should permit observation and range determination to 800 meters.

b. The ideal field firing range should be located on terrain that has been left primarily in its natural state. The range should be a minimum of 800 meters in depth with provisions along the firing line for several sniper positions within each lane to provide a slightly different perspective of the target area. Where time precludes construction of a separate range, it may be necessary to superimpose this facility over an existing field firing range. Ideally, the field firing range would have targets emplaced the following way:

Meters	Type Target
200	E-type silhouette, hit-kill mechanism.
300	Iron maiden silhouette; E-type silhouette, hit-kill mechanism; moving target mechanism.
325	E-type silhouette, hit-kill mechanism.
375	E-type silhouette, hit-kill mechanism, emplaced inside a window.
400	E-type silhouette, hit-kill mechanism, emplaced inside a bunker.
500	Iron maiden silhouette; moving target mechanism, tracked vehicle with a hit-kill mechanism in the commander's cupola.
600-1,000	Iron maiden silhouettes.

(1) Iron maidens can be made out of 3/4-inch steel plate with a supporting frame. It should be cut out in the form of a silhouette 20 inches wide and 40 inches high. By painting these targets white, snipers can easily detect where the bullet impacted on the target.

(2) Placing targets inside of window openings gives the sniper experience engaging targets that can be found in an urban environment. This can be done by cutting a 36- by 48-inch hole in the center of an 8- by 16-foot plywood wall and emplacing an E-type silhouette on a hit-kill mechanism 2 to 4 meters behind the wall.

(3) Placing targets inside of a bunker type of position allows the sniper to gain experience in firing into darkened openings. This position can be built with logs and sandbags with an E-type silhouette on a hit-kill mechanism placed inside.

(4) Moving targets may be used between 300 and 500 meters to give the sniper practical experience and development of proficiency in engaging a moving target. Two targets, one moving laterally and one moving at an oblique, will present a challenge to the sniper.

(5) Targets should be arranged to provide varying degrees of concealment to depict enemy personnel or situations in logical locations. The grouping of two or more targets to indicate a crew-served weapon situation or a small unit is acceptable. Such arrangements, provided the targets can be marked, may require selective engagement by the sniper. The automatic target devices provide for efficient range operation and scoring. Figure A-1 shows the target layout for one lane.

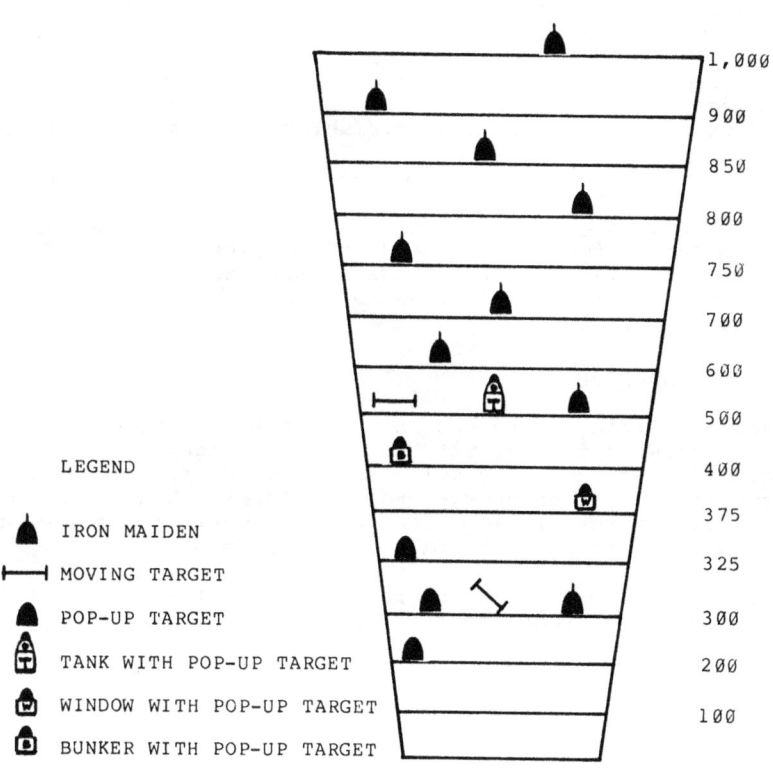

Figure A-1. Lane layout.

    c. Concealment and concealed movement areas should be in natural terrain that allows observation of the entire area out to 1,000 meters.

    d. Observation exercises can be conducted in a 100- by 100-meter cleared area with one side bordered by a tree line or a hill to serve as a forward limit for observing.

    e. Range estimation exercises should be conducted in an area that provides good observation out to 1,000 meters.

A-2. ZEROING AND PRACTICE FIRE

To engage targets effectively during training exercises and in combat, the sniper must have his rifle zeroed accurately. For this reason, the zeroing exercises are normally conducted on a measured known-distance range to ensure precise adjustment, recording, and practice under ideal conditions and to eliminate variables that may detract from achieving an effective zero. The sniper rifle is zeroed using both the iron and scope sights. An international bull-type target should be used for zeroing. It is important to acquire a point-of-aim -- point-of-impact zero at 300 meters for the M21 and 100 meters for the M24. As the distance increases, the sniper must adjust his scope to allow for elevation and wind to ensure his rounds stay in the center circle.

A-3. FIELD FIRE

Practical firing exercises are designed to develop sniper proficiency in the accurate and rapid engagement of various combat-type targets, as well as to provide practical work in other field techniques. Sniper teams should be given positions on the firing line and areas of the field fire course to observe and to make range cards of the area.

    a. After the range cards have been completed, the sniper team will be required to fire the course by having one member call the wind and adjust the other member's fire. It is just as important to be able to call the wind as it is to successfully engage the targets. After one member fires the course, they switch over and repeat the fire course.

    b. When firing the course, snipers should engage the targets in a sequence that starts at the 200-meter target, then engage each target out to 800 meters, then engage targets back to the 200-meter target. The course, consisting of the engagement of 20 targets and 30 rounds of ammunition (target can only be engaged twice), should be scored as follows:

o   10 points for first-round hits.

o   5 points for second-round hits.

o   200 points maximum.

o   160 points needed to pass.

c. To enhance training, the field fire course should also be fired during limited visibility with overhead illumination, such as parachute flares. This puts stress on the sniper to determine the range and engage a target in a short amount of time.

d. In order to provide the most realistic training environment, range commands to commence fire and cease fire are not used in sniper exercises. The only exception to this is in the event of an unsafe condition. In that case, the command CEASE FIRE should be given immediately. Snipers must be given a thorough orientation on each exercise (to include safety requirements) before they are permitted to move into position. Anytime after the sniper has assumed his firing position in the designated location, he should be allowed to fire without further commands. Therefore, the range must be cleared for firing before the exercise begins. An NCO (assistant instructor) must be with each sniper to keep score and ensure safety is maintained during the exercise. When the sniper completes firing, the NCO ensures the rifle is clear and signals the range officer that the weapon is clear.

A-4. CONCEALMENT

Concealment exercises are designed to develop and test the sniper's ability to conceal himself in an expedient firing position to observe and engage an observer.

a. In a cleared area with a wood line approximately 100 meters away, give the snipers 10 minutes to conceal themselves in the wood line. After the 10-minute preparation period is over, have an observer with binoculars and an M49 telescope observe the wood line from 100 meters away for 20 minutes. If there are more than 10 snipers in the exercise, two observers and two walkers may be needed. After the 20-minute period, have a walker with a radio move within 10 feet of a sniper who is ready to fire at an observer.

b. The sniper should be able to identify a white 5-inch number painted on an 8-inch by 8-inch international orange panel that is held over a vital part of the observer's body and fire two blanks at him without being detected. If the observer detects the sniper, he will radio the walker and direct him to the sniper. The exercise should be scored on a 1- to 10-point system, with 7 being a passing score.

IF THE SNIPER HAS --	POINT VALUE 10
Been detected in the first 2 minutes of the exercise.	2
Been detected with the aid of optics.	3
(Walker will now find a sniper who is able to engage an observer.)	
Not been detected when walker is within 10 feet of shooter.	4
Properly identified number within 30 seconds.	5
Failed to properly identify number.	2
Fired first shot, not detected.	6
Fired second shot, not detected.	8
Stable firing position with support.	9
Weapon's scope properly adjusted for range and windage.	10

NOTE 1: If sniper is caught trying to identify number, he will receive four points.

NOTE 2: If muzzle blast/flash is detected, one point will be deducted from total score.

NOTE 3: Failing to comply with training standards and objectives (unnecessary movement, premature fire, outside of prescribed boundaries) will result in termination of the exercise and a score of zero.

A-5. CONCEALED MOVEMENT

The purpose of the concealed movement exercise is to develop and test the sniper's ability to move and occupy a firing position undetected.

    a. This exercise requires the same amount of instructors and equipment as used in the concealment exercises. Areas used for this exercise should be observable for 1,000 meters and have easily recognizable left and right limits. Ideally, snipers should train in a different type of area each time they participate in these exercises.

    b. The snipers are to move 600 to 800 meters toward two observers, occupy a firing position 100 to 200 meters away, identify in the same manner as the concealment exercise, and fire two blanks at the observers without being detected any time during the exercise. If one of the observers detects a sniper, he will radio one of the walkers and direct him to the sniper's position. The exercise is scored on a 1- to 10-point system, with 7 being a passing score.

IF THE SNIPER HAS --	POINT VALUE 10
Been detected moving to the FFP.	2
Been detected moving in the FFP.	3
Fired first shot, not detected.	4
Not been detected when walker is within 10 feet of shooter.	5
Properly identified number within 30 seconds.	6
Failed to properly identify number.	3
Not been detected when walker is within 5 feet of shooter.	7
Fired second shot, not detected.	8
Stable firing position with support.	9
Weapon's scope properly adjusted for range and windage.	10

NOTE 1: If muzzle blast/flash is detected, one point will be deducted from total score.

NOTE 2: Failing to comply with training standards and objectives (unnecessary movement, premature fire, outside of prescribed boundaries) will result in termination of the exercise and a score of zero.

A-6. TARGET DETECTION

Target detection exercises sharpen the sniper's eyes by requiring him to detect, describe, and plot objects that cannot be easily seen or described without skillful use of optics.

   a. Areas used for target detection should be partially cleared at least 200 meters in depth and 100 meters in width with easily definable left and right limits. The area should have at least three target reference points that are easily recognized and positioned in different locations throughout the area. Ten military items are placed in the area. These items can be radio antennae, small-scale mock vehicles, batteries, map protractors, or weapons. Items should be placed so that they are undetectable with the naked eye, detectable but indescribable with the binoculars, and describable only by using the M49 telescope.

   b. Snipers are given 40 minutes to detect, describe, and plot each item in the area. Snipers will remain in the prone position throughout the exercise. After the first 15 minutes, they will move to a different position, left or right of the center line of observation and remain there for the next 15 minutes. For the last 10 minutes, they can choose a position anywhere along the line for the remainder of the exercise. When an object is detected, the sniper must plot its location on a prepared scorecard (Figure A-2) by giving a mil reading and direction (clock method) from a TRP and the sniper's location on the line of observation (left or right). Next, the sniper must describe the object using the categories of size, shape, color, condition, and appears to be. Snipers will receive 1/2 point for correctly plotting a target and 1/2 point for correctly describing it. They must achieve seven points to receive a go in this area.

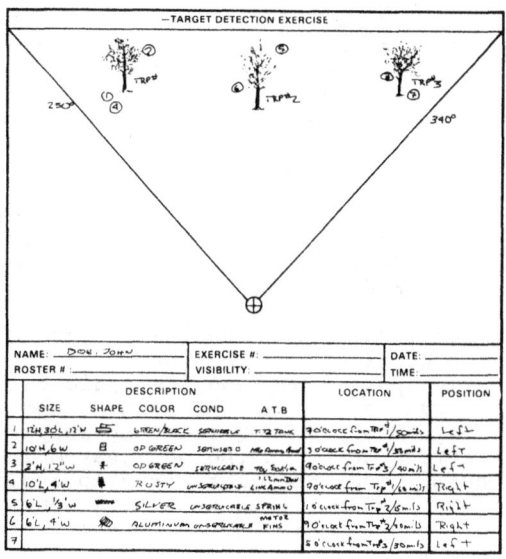

Figure A-2. Target detection scorecard.

A-7. RANGE ESTIMATION

Snipers must be able to correctly estimate distance in order to fire their weapons effectively, make accurate range cards, and give reliable intelligence reports. Range estimation exercises should be conducted in an area that allows unobstructed observation of a human-size target up to 1,000 meters away. Targets such as vehicles or personnel should be placed at various ranges and stages of concealment to give the sniper a challenging and realistic exercise. The snipers should be graded on their ability to estimate range by using the naked eye, M19 binoculars, or the rifle scope. Snipers must correctly estimate the distance to 7 of 10 objects using their eyes (+/- 12 percent) and correctly estimate the distance of 7 to 10 objects using optics, binoculars (+/- 10 percent), and an M3A scope (+/- 5 percent).

A-8. LAND NAVIGATION

This exercise is designed to develop sniper team proficiency in specific field techniques, such as movement, land navigation, and radiotelephone procedure. The team will be required to move from a starting point to a specific location and then report. During this exercise, the sniper team should be equipped as indicated in Chapter 2. To provide training under varied conditions, this team exercise should be conducted at least two times, once during daylight and once after nightfall.

   a. This exercise can be held at the same time as the team firing exercises. Half of the training class or group could conduct the land navigation exercise, while the other half conducts the firing exercise. When they finish, they change over.

   b. The sniper teams are assembled at the starting point and given instructions indicating the mission objective, the observation positions, and the radio call signs. They are also given an equipment check and an exercise briefing. This exercise requires the sniper team to move from the starting point to the designated location in less than two hours. They are instructed to avoid the observation positions, which represent the enemy. They are required to report their location every 15 minutes and their arrival at the destination site. A starting team is given a score of 100 points and the following deductions are made for errors.

   o 10 points off for each time the sniper team is seen by someone in the observation positions.

   o 5 points off for each instance of improper radio procedure or reporting.

   o 1 point off for each minute over the authorized two hours.

   o 3 points off for every 5 meters that the team misses the designated destination.

   o 100 points off for being lost and failing to complete the exercise.

   d. At the completion of this exercise, the instructor will critique the sniper teams on their performance.

A-9. "KEEP IN MEMORY" GAMES

A sniper must have a good memory to report facts accurately even though he may not have had the time or means to record them at the time he saw them.

a. KIM games exercise the sniper's memory capacity by requiring him to remember items he has viewed and recall them later. This is done by placing 10 items on a table or on the ground and covering them with a towel, poncho, or anything suitable.

b. The sniper is given two minutes to view the items without touching or moving them. He will then be given two minutes to record what he has seen by using the five categories of size, shape, color, condition, and appears to be. Exercises can be varied to give the sniper more of a challenge. Examples of this are to shorten the amount of time to view/write, to create distractions while viewing/writing, to lengthen the time between viewing and writing, and to have the sniper conduct some type of activity between viewing and writing. Practicing drill and ceremony works quite well, because this requires the sniper to concentrate on marching movements and not on what he has viewed.

A-10. RECORD EXERCISES

The true test of the sniper or sniper team is the successful accomplishment of the assigned mission. Realizing that the employment possibilities for the sniper are broad, it is difficult to test them in all skill requirements. Their most important skill is the ability to use one shot to hit (kill) one target. The exercises outlined in the previous paragraphs may be executed once for practice and once for record to determine an individual's sniper qualification score. These exercises are, collectively, the best practical way in peacetime to evaluate the sniper's capability to function.

TC 23-14

APPENDIX B
SNIPER SUSTAINMENT PROGRAM

Sniper skills are very perishable and must be exercised as often as possible. A sniper sustainment program will provide the individual sniper or team an opportunity to sustain their skills in both marksmanship and fieldcraft on a regular basis. Units should try to develop a sustainment program that requires the sniper teams to requalify with their sniper weapons on a quarterly basis. In conjunction with the marksmanship training, the sniper would be required to participate in fieldcraft training. Training should be at least one week long. Coupled with the quarterly refresher training, units should plan a sniper emergency deployment readiness exercise, which would be 24 hours long. At a minimum, the EDRE would require the sniper to engage targets with the sniper weapon on a field-fire range, participate in a concealed movement exercise, and negotiate a day/night land navigation course.

B-1. TRAINING

An example of a five-day sniper sustainment program is as follows:

-- DAY 1 --

TASK: Select sniper team routes and positions.

CONDITIONS: Given a review of selection of routes and positions, a situational sniper mission with a target area location that requires a minimum movement of 3,000 meters, a military map, a protractor, a felt-tip pen, an 8-inch square clear plastic overlay, and one sheet of letter-size paper.

STANDARDS: Select and plot a primary and alternate route, objective rally point, and tentative sniper firing position that provides the best cover and concealment.

1. Prepare overlay with two grid reference marks; primary and alternate routes with arrows indicating direction of travel; minimum of three checkpoints, numbered in order; objective rally point; and a tentative firing position.

2. Prepare a written log of movement using the sniper patrol order format (paragraph IIIc(3)). Log will contain the from/to grid coordinates, magnetic azimuths, distance, checkpoint number, objective rally point, and tentative firing position.

3. Prepare overlay and written log of movement within 30 minutes.

TASK: Move while using individual sniper movement techniques.

CONDITIONS: Given a review of sniper movement techniques, a sniper weapon, a ghillie suit, and a flat, open area that allows trainers to observe movement techniques.

STANDARDS: Move correctly while using the designated movement technique.

1. Sniper low crawl.
2. Medium crawl.
3. High crawl.
4. Hand-and-knee crawl.
5. Walking.

NOTE: Trainers will designate movement techniques and critique snipers on their movement.

TASK: React to enemy contact while moving as a member of a sniper team.

CONDITIONS: Given a review of sniper team movement techniques and reactions to enemy contact, sniper team's basic equipment/weapons, and an area of varying terrain with at least one danger area.

STANDARDS: React correctly to designated situations or danger areas.

1. Visual contact.
2. Ambush.
3. Indirect fire.
4. Air attack.
5. Danger area (linear and open area).

NOTE: Trainers will designate situations and critique sniper teams on their movement.

TASK: Describe target detection, selection, and observation techniques.

CONDITIONS: Given a review of target detection, selection, and observation techniques.

STANDARDS: Describe, orally or in writing, techniques used to observe, detect, and select targets.

TASK: Identify Threat uniforms, equipment, and vehicles.

CONDITIONS: Given a review of pictures or slides of Threat uniforms, equipment, and vehicles.

STANDARDS: Identify 7 of 10 Threat uniforms or rank insignia, 7 of 10 pieces of Threat equipment, and 7 of 10 Threat vehicles.

TASK: Describe range estimation techniques.

CONDITIONS: Given a review of range estimation techniques used by snipers.

STANDARDS: Describe, orally or in writing, range estimation techniques used by the sniper.

1. Eye methods.

2. Use of binoculars.

3. Use of sniperscopes.

TASK: Prepare a sniper range card.

CONDITIONS: Given a review of sniper range cards, a suitable target area, basic sniper equipment, and a sniper range card.

STANDARDS: Prepare a sniper range card complete with --

1. Grid coordinates of position.

2. Target reference point(s) (azimuth, distance, and description).

3. Left/right limits with azimuths.

4. Ranges throughout area.

5. Major terrain features.

6. Method of obtaining range/name.

7. Weather data.

TASK: Prepare a military sketch.

CONDITIONS: Given a review of sniper military sketching, a suitable area or object to sketch, and a blank military sketch sheet.

STANDARDS: Prepare a sketch complete with --

1. Grid coordinates of position.

2. Magnetic azimuth through center of sketch.

3. Sketch name and number.

4. Scale of sketch.

5. Remarks section.

6. Name/rank.

7. Date/time.

8. Weather data.

TASK: Maintain a sniper log.

CONDITIONS: Given a review of sniper logs and 20 blank sheets stapled together as a booklet.

STANDARDS: Maintain a sniper log with chronological listing of events that take place during the next three days and containing the following:

1. Grid coordinates of position.

2. Observer's name.

3. Date/time/visibility.

4. Sheet number/number of total sheets.

5. Series number/time and grid coordinate of each event.

6. Event.

7. Action taken.

NOTE: Trainers will collect the logbooks in three days.

-- DAY 2 --

TASK: Describe the fundamentals of sniper marksmanship.

CONDITIONS: Given a review of sniper marksmanship fundamentals.

STANDARDS: Describe, orally or in writing, the fundamentals of sniper marksmanship.

1. Position.
2. Breath control.
3. Aiming.
4. Trigger control.

TASK: Describe the effects of weather on ballistics.

CONDITIONS: Given a review of the effects of weather on ballistics.

STANDARDS: Describe, orally or in writing, the effects of weather on ballistics.

TASK: Describe the sniper team method of engaging targets.

CONDITIONS: Given a review of the sniper team method of engaging targets.

STANDARDS: Describe, orally or in writing, the sniper team method of engaging targets.

TASK: Describe methods used to engage moving targets.

CONDITIONS: Given a review of methods used to engage moving targets.

STANDARDS: Describe, orally or in writing, methods used to engage moving targets.

TASK: Describe methods used to engage targets at various ranges without adjusting the scope's elevation.

CONDITIONS: Given a review of methods used to engage targets at various ranges without adjusting the scope's elevation.

STANDARDS: Describe, orally or in writing, the methods used to engage targets at various ranges without adjusting the scope's elevation.

TASK: Zero metallic sights.

CONDITIONS: Given a sniper weapon, a suitable firing range, and 12 rounds of 7.62-mm special ball ammunition.

STANDARDS: Zero metallic sights on a sniper weapon within 12 rounds.

TASK: Participate in a field fire exercise using metallic sights.

CONDITIONS: Given a sniper weapon, M49 telescope, a suitable firing range, and 20 rounds of 7.62-mm special ball ammunition.

STANDARDS: Engage targets from 200 to 700 meters achieving a minimum of 16 hits, using metallic sights.

-- DAY 3 --

TASK: Zero rifle scope.

CONDITIONS: Given a sniper weapon, an M49 telescope, a suitable firing range, and 12 rounds of 7.62-mm special ball ammunition.

STANDARDS: Zero rifle scope within 12 rounds.

TASK: Engage moving targets.

CONDITIONS: Given a sniper weapon, an M49 telescope, a suitable firing range, and 10 rounds of 7.62-mm special ball ammunition.

STANDARDS: Engage 10 moving targets, from 300 to 500 meters, achieving a minimum of 8 hits.

TASK: Estimate range.

CONDITIONS: Given a sniper weapon (M24), M19 binoculars, and 10 targets out to 800 meters.

STANDARDS: Correctly estimate range to 7 of the 10 targets using eye estimation (+/- 12 percent), binoculars (+/- 10 percent), or the M24 sniper weapon (+/- 5 percent).

TASK: Detect targets.

CONDITIONS: Given a suitable area with 10 military objects, binoculars, M49 telescope, and a score sheet.

STANDARDS: Detect, plot, and describe 7 of 10 military objects within 40 minutes.

TASK: Participate in a concealment exercise.

CONDITIONS: Given a sniper weapon, ghillie suit, two 7.62-mm blank rounds of ammunition, an area to conceal a sniper position, and 10 minutes to prepare.

STANDARDS: Without being detected, occupy a position, identify, and fire two blank rounds at an observer (located 100 to 200 meters away) who is equipped with binoculars and an M49 telescope. Must score 7 of 10 points. Points are as follows:

If the sniper -- Points

   Has been detected without the aid of optics
   (first 2 minutes), score 2

Has been detected with the aid of optics
(18 minutes), score 3

Has not been detected when walker is within
10 feet of shooter, score 4

Has properly identified number within 30
seconds, score 5

Has failed to properly identify number, score 2

Has fired first shot, not detected, score 6

Has fired second shot, not detected, score 7

Has stable firing position (support), score 9

Has properly adjusted weapon's scope for
range and windage, score 10

NOTE 1: If sniper is caught trying to identify number, he will receive 4 points.

NOTE 2: If muzzle blast/flash is detected, deduct 1 point from total score.

NOTE 3: Failing to comply with training standards and objectives (such as unnecessary movement, premature fire, outside of prescribed boundaries) will result in termination of the exercise and a score of zero.

-- NIGHT 3 --

TASK: Engage targets during darkness.

CONDITIONS: Given a sniper weapon, suitable firing range, four overhead flares, and 12 rounds of 7.62-mm special ball ammunition.

STANDARDS: Engage three targets from 200 to 500 meters within the time the target area is illuminated.

NOTE 1: Trainers will fire overhead flares one at a time.

NOTE 2: Snipers will fire four separate times.

-- DAY 4 --

TASK: Participate in a field fire exercise.

CONDITIONS: Given a sniper weapon, M49 telescope, a suitable firing range, and 20 rounds of 7.62-mm special ball ammunition.

STANDARDS: Engage targets from 200 to 900 meters, achieving at least 16 hits.

TASK: Participate in a concealed movement exercise.

CONDITIONS: Given a sniper weapon, ghillie suit, two 7.62-mm blank rounds of ammunition, and a suitable area 1,000 meters long that is observable.

STANDARDS: Move 600 to 800 meters; without being detected, occupy a position, identify, and fire two blank rounds at an observer who is equipped with binoculars and an M49 telescope within four hours. Must score 7 of 10 points. Points are as follows:

If the sniper -- Points

   Has been detected moving to the FFP, score 2

   Has been detected moving in the FFP, score 3

   Has fired first shot, not detected, score 4

   Has not been detected when walker is within
   10 feet of shooter, score 5

   Has properly identified number (within 30
   seconds), score 6

   Has failed to properly identify number, score 3

   Has not been detected when walker is within
   5 feet of shooter, score 7

   Has fired second shot, not detected, score 8

   Has stable firing position (support), score 9

   Has properly adjusted weapon's scope for
   range and windage, score 10

NOTE 1: If muzzle blast/flash is detected, deduct 1 point from the total score.

NOTE 2: Failing to comply with training standards and objectives (such as unnecessary movement, premature fire, outside of prescribed boundaries) will result in termination of the exercise and a score of zero.

-- DAY 5 --

TASK: Call for fire.

CONDITIONS: Given a review of call for fire procedures, two AN/PRC-77 radios, and a fire mission.

STANDARDS: Transmit the fire mission using proper radio procedures and the elements of the call for fire mission in sequence:

1. Observer identification.
2. Warning order.
3. Target location.
4. Target description.
5. Method of engagement (optional).
6. Method of fire and control (optional).

TASK: Locate target by grid coordinates.

CONDITIONS: Given a review of locating targets using the grid coordinate method, a map of the target area, binoculars, compass, and a target.

STANDARDS: Determine and announce the six-digit coordinates of the target (within a 250-meter tolerance) within 30 seconds.

TASK: Locate a target by polar plot.

CONDITIONS: Given a review of target locating using the polar plot method, a map of the target area, binoculars, a compass, and a target.

STANDARDS: Locate the target within 250 meters of the actual location. Announce the target location within 30 seconds after identification. Express direction to the

B-10

nearest 10 mils and within 100 mils of actual direction. Express distance to the nearest 100 meters.

TASK: Locate target by shift from a known point.

CONDITIONS: Given a review of locating targets using the shift from a known point method, a map of the target area, binoculars, a compass, a known point, and a target.

STANDARDS: Locate the target within 250 meters of the actual location and announce the target location within 30 seconds after identification. Express direction to the nearest 10 mils and within 100 mils of the actual direction. Express right or left corrections to the nearest 10 meters and range corrections to the nearest 100 meters.

TASK: Participate in a land navigation exercise during daylight.

CONDITIONS: Given a navigation course with at least four legs no less than 800 meters apart.

STANDARDS: Navigate the course without being detected. Preparing sketches, range cards, and/or logs can also be incorporated into the exercise.

-- NIGHT 5 --

TASK: Participate in a land navigation exercise during nightfall.

CONDITIONS: Given a navigation course with at least three legs no less than 500 meters apart. Observers can be placed on the course to detect any violations of noise and light discipline and deduct points from the sniper's score for violations.

STANDARDS: Navigate the course without being detected.

NOTE: Refer to FM 21-26 for guidance in preparing a land navigation course and requirement sheets.

B-2. EMERGENCY DEPLOYMENT READINESS EXERCISE

An example of a battalion EDRE is as follows:

TIME	ACTION
0400	Battalion alerts company sniper teams.

    1. CQ relays uniform and packing list.

    2. Sniper teams have two hours to report to company.

    3. Sniper team leaders report to SEO when all of the team is accounted for.

    4. Sniper team receives FRAGO from the SEO.

0600   Snipers depart company area by air, truck, or road march.

0800   Sniper teams arrive at range.

    1. Sniper teams receive range/safety briefing.

    2. Snipers receive issued ammunition.

    3. Snipers zero weapons.

    4. Sniper teams field/record on a range with targets positioned from 200 to 900 meters.

1100   Sniper teams depart range; move to concealed movement site by truck, road march, or tactical movement by teams.

1200   Sniper teams arrive at concealed movement site.

    1. Sniper teams receive briefing.

    2. Site should be 800 to 1,000 meters long. Trained observer should be positioned at one end with field table, M19 binoculars, M49 spotting scope, 8-inch by 8-inch international orange panels with white 5-inch number (1 to 9) painted on them, PRC-77 radio for him and walker.

TIME	ACTION
	3. Sniper will have four hours to move into his FFP, 50 to 200 meters from observer, and fire his first shot.
	4. Sniper will have 30 seconds in which to identify number.
	5. Sniper will fire second shot.
	6. The entire exercise will be conducted without the sniper being detected by the observer.
1600	Sniper teams depart for day/night land navigation exercise.
	1. Sniper teams start exercise from concealed movement site.
	2. Sniper teams will be required to move to three different points. At each point, they will perform one of the following: o Draw a military sketch. o Draw a range card. o Do a target detection exercise. o Collect information/data.

NOTE: All information is to be recorded in the sniper logbook.

	3. All movement will be performed without being detected.
2000	Night navigation exercise.
	1. Sniper teams start exercise from CP.
	2. They will move undetected to three different points.
	3. They will perform a detection exercise with the use of NODs.

B-13

TIME	ACTION
	4. They will record all information in the sniper logbook.
	5. After collecting necessary data, they will move to an extraction point and construct a sniper hide position. They will prepare for target reduction.
0500-0600	Target reduction.
	1. Upon target reduction time, the team will prepare for extraction.
	2. At extraction time, they will return to the company area.
	3. The team will debrief the sniper employment officer.
	4. The SEO will conduct an after-action review.

NOTE: A written test could also be given as part of the EDRE.

TC 23-14

APPENDIX C
SNIPER'S DATA CARD

The sniper's data card is prepared by the sniper to record the results and all elements that had an effect on the firing of the weapon (Figure C-1). This can vary from information about weather conditions to the attitude of the firer on that particular day. The sniper can refer to this information in the future to understand his weapon, the weather effects, and his shooting ability on a given day. One of the most important items of information he will record is the cold barrel zero of his weapon. There are three phases in writing information on the data card. These are before firing, during firing, and after firing.

C-1. BEFORE FIRING

Information that is written before firing is:

o Range. The distance to the target.

o Rifle and scope number. The serial numbers of the rifle and scope.

o Date. Date of firing.

o Ammo. Type and lot number of ammunition.

o Light. Amount of light (overcast, clear, and so forth).

o Mirage. Can a mirage be seen (good, bad, fair, and so forth)?

o Temp. Temperature on the range.

o Hour. Time of firing.

o Light (diagram). Draw an arrow in the direction the light is shining.

o Wind. Draw an arrow in the direction the wind is blowing, and record its average velocity and cardinal direction (N, NE, S, SW, and so forth).

C-2. DURING FIRING

Information that is written while firing is:

a. Elevation. Elevation setting used and any correction needed. Example: Target distance is 600 meters; sniper sets elevation dial to 6. Sniper fires and round hits target 6 inches low of center. He then adds one minute (one click) of elevation (+1).

b. Windage. Windage setting used and any correction needed. Example: Sniper fires at 600-meter target with windage setting on 0; round impacts 15 inches right of center. He will then add 2 1/2 minutes left to the windage dial (L/2 1/2). When firing the M21, the sniper draws the windage holdoff on the silhouette in the "HOLD" box.

c. Shot. The column of information about a particular shot. Example: Column 1 is for the first round, column 10 is for the tenth round.

d. Elev. Elevation used. Example: (6+1), (6), (6-1).

e. Wind. Windage used. Example: (L/2 1/2), (0), (R/1/2).

f. Call. Where the aiming point was when the weapon fired.

g. Large Silhouette. The large silhouette is used to record the exact impact of the round on the target. This is recorded by writing the shot's number on the large silhouette in the same place it hit the target.

C-3. AFTER FIRING

After firing, the sniper will write any comments about firing in the remarks section. This can be comments about the weapon, firing conditions (time allowed for fire), or his condition (nervous, felt bad, felt good, and so forth).

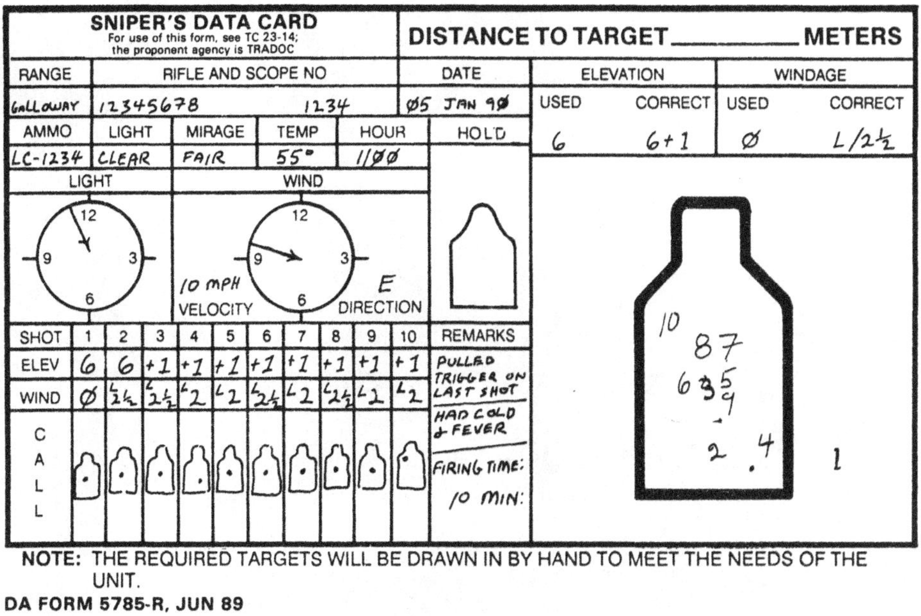

SNIPER'S DATA CARD For use of this form, see TC 23-14; the proponent agency is TRADOC				DISTANCE TO TARGET _____ METERS				
RANGE	RIFLE AND SCOPE NO			DATE	ELEVATION		WINDAGE	
GALLOWAY	12345678		1234	Ø5 JAN 9Ø	USED	CORRECT	USED	CORRECT
AMMO	LIGHT	MIRAGE	TEMP	HOUR	HOLD			
LC-1234	CLEAR	FAIR	55°	11ØØ	6	6+1	Ø	L/2½

LIGHT — 10 mPH VELOCITY — E DIRECTION (WIND)

SHOT	1	2	3	4	5	6	7	8	9	10	REMARKS
ELEV	6	6	+1	+1	+1	+1	+1	+1	+1	+1	PULLED TRIGGER ON LAST SHOT
WIND	Ø	L2½	L2½	L2	L2	L2½	L2	L2½	L2	L2	HAD COLD & FEVER
C A L L											FIRING TIME: 10 MIN

NOTE: THE REQUIRED TARGETS WILL BE DRAWN IN BY HAND TO MEET THE NEEDS OF THE UNIT.

DA FORM 5785-R, JUN 89

Figure C-1. Prepared sniper's data card.

C-3

TC 23-14

APPENDIX D
MEASUREMENTS

A sniper must have a good working knowledge of the angular measurements involved with shooting. These measurements aid the sniper in estimating distances and adjusting for bullet trajectories. The two most common types of measurements the sniper uses are mils and minutes of angle.

D-1. MILS

A mil is an angular measurement that is equal to 1/1,000th the distance to an object. For example, the apparent size of a 1-meter target 1,000 meters away is 1 mil. At 500 meters a 1-meter target appears to be 2 mils (Figure D-1). Mils are used mostly for range estimations (see Chapter 4), but can also be used for elevation/windage holdoffs and moving target leads when using the M3A scope (see Chapter 3).

D-2. MINUTE OF ANGLE

A minute of angle is 1/60th of a degree (Figure D-2). This equals approximately 1 inch (1.145 inches) for every 100 meters. Examples: 1 MOA = 2 inches at 200 meters; 1 MOA = 5 inches at 500 meters. Minutes of angle are used to determine and adjust the elevation and windage needed on the weapon's scope.

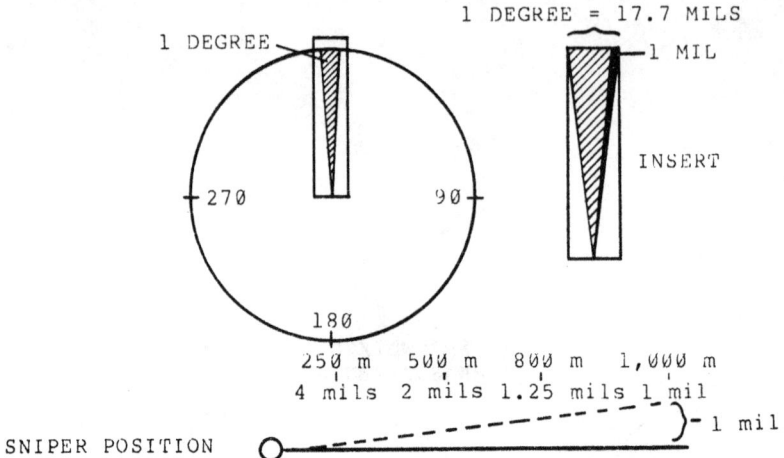

Figure D-1. Mils.

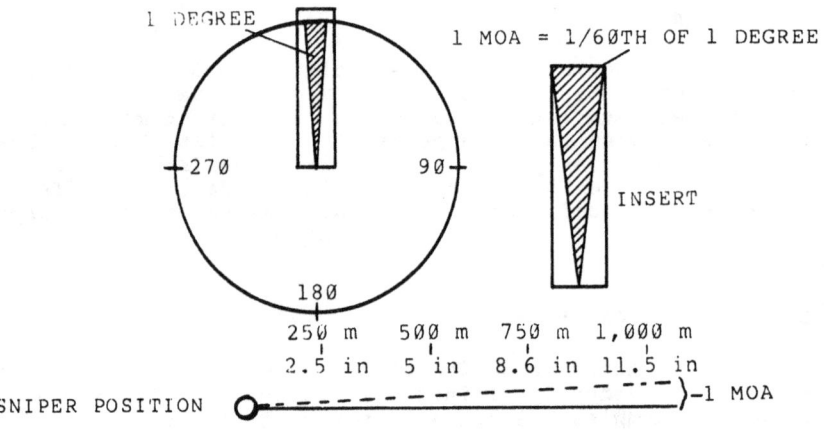

Figure D-2. Minutes of angle.

TC 23-14

APPENDIX E
REFERENCE TABLES

This appendix provides a ready reference when determining effects of wind on ballistics, ballistic trajectories, and estimating range.

RANGE Meters	WIND Value	3 mph min	3 mph in	5 mph min	5 mph in	7 mph min	7 mph in	10 mph min	10 mph in	12 mph min	12 mph in	15 mph min	15 mph in	18 mph min	18 mph in	20 mph min	20 mph in
200	Half	0	.4	.5	.6	.5	.8	.5	1.2	.5	1.3	1	1.8	1	2.2	1	2.4
	Full	.5	.8	.5	1.2	1	1.7	1	2.4	1.5	2.9	1.5	3.6	2	4.3	2	4.8
300	Half	.5	.9	.5	1.3	.5	1.9	1	2.7	1.5	3.3	1.5	4	1.5	4.9	1.5	5.4
	Full	.5	1.7	1	2.7	1	3.8	1.5	5.4	2	6.5	2.5	8.1	3	9.8	3.5	10.9
400	Half	.5	1.4	.5	2.4	1	3.3	1.5	4.8	1.5	5.8	1.5	7.2	2	8.6	2	9.6
	Full	.5	2.9	1	4.8	1.5	6.7	2	9.6	2.5	11.5	3.5	14.4	4	17.3	4.5	19.2
500	Half	.5	2.3	1	3.8	1	5.3	1.5	7.5	1.5	9	2	11.3	2.5	13.5	2.5	15
	Full	1	4.5	1.5	7.5	2	10.5	2.5	15	3.5	18	4	22.6	5	27	5.5	30
600	Half	.5	3	1	5	1.5	8	1.5	11	2	13	2.5	16	3	19	3	22
	Full	1	7	1.5	11	2.5	15	3	21	4	26	5	32	6	39	6.5	43
700	Half	1	4	1	7	1.5	10	1.5	15	2.5	18	3	22	3.5	26	4	29
	Full	1.5	9	2	15	3	21	3	29	4.5	35	6	44	7	53	7.5	59
800	Half	1	6	1	10	1.5	13	2	19	2.5	23	3.5	29	4	35	4.5	38
	Full	1.5	11	2	18	3	27	4	38	5.5	46	5.5	57	8	69	7.5	77
900	Half	1	7	1.5	12	1.5	17	2.5	24	3	29	3.5	36	4.5	44	5	49
	Full	1.5	15	2.5	24	3.5	34	4.5	49	6	58	7.5	73	9	97	10	97
1,000	Half	1	9	1.5	15	2	21	2.5	30	3.5	36	4	45	5	54	5.5	60
	Full	1.5	18	2.5	30	4	42	5.5	60	6.5	72	8	90	10	103	11	120

NOTE: For ball ammunition, increase holdoff measurement by 13 percent.

Figure E-1. Wind conversion table.

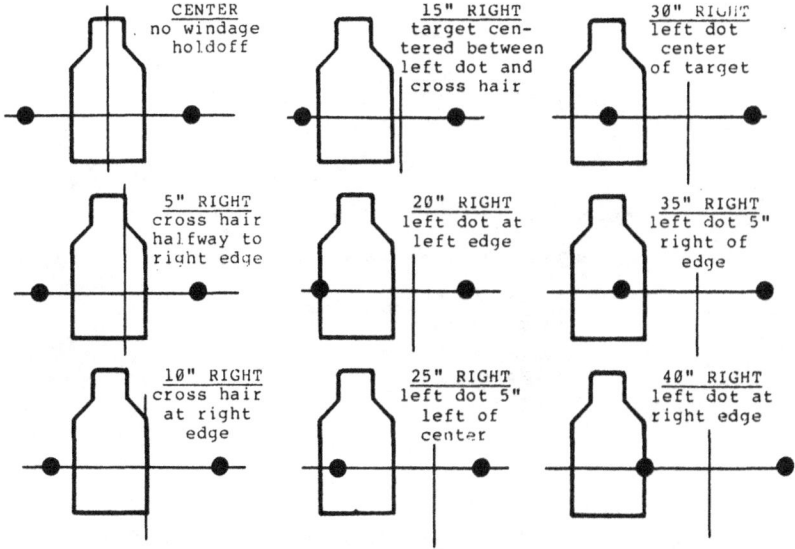

Figure E-2. Holdoff for 7.62 (173 grains) in inches (M21 system).

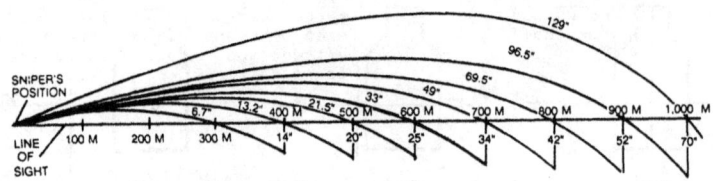

Figure E-3. Trajectory chart.

TABLE FOR 6' MAN

HEIGHT IN MILS	STANDING	SITTING/KNEELING
1	2000	1000
1.5	1333	666
2	1000	500
2.5	800	400
3	666	333
3.5	571	286
4	500	250
4.5	444	222
5	400	200
5.5	364	182
6	333	167
6.5	308	154
7	286	143
7.5	267	133
8	250	125
8.5	235	118
9	222	111
9.5	211	105
10	200	100

TABLE FOR 5' 6" MAN

HEIGHT IN MILS	STANDING	SITTING/KNEELING
1	1800	900
1.5	1200	600
2	900	450
2.5	750	375
3	600	300
3.5	514	257
4	450	225
4.5	400	200
5	360	180
5.5	327	164
6	300	150
6.5	277	139
7	257	129
8	225	113
9	200	100
10	180	90

RANGE ESTIMATION TABLE OF MILS FOR OBJECTS

FEET	3	4	5	6	7	8	9	10	11	12	13	14	15	16	17	18
YARDS	1	1.3	1.7	2	2.3	2.7	3	3.3	3.7	4	4.3	4.7	5	5.3	5.7	6
2	500	650	850	1000	1150	1350	1500	1650	1850	2000	2150	2350	2500	2650	2850	3000
2.5	400	520	680	800	920	1080	1200	1320	1480	1600	1720	1880	2000	2120	2280	2400
3	333	425	566	666	766	900	999	1100	1230	1332	1433	1566	1665	1766	1900	1998
3.5	285	371	486	571	657	771	855	943	1057	1140	1229	1343	1425	1514	1629	1710
4	250	325	425	500	575	675	750	825	925	1000	1075	1175	1250	1325	1425	1500
4.5	222	289	378	444	511	600	666	733	822	888	950	1044	1110	1178	1267	1332
5	200	260	340	400	460	540	600	660	740	800	860	940	1000	1060	1140	1200
5.5	182	236	309	362	418	491	543	600	673	724	782	855	905	964	1036	1086
6	167	217	283	334	383	450	500	550	617	668	717	783	835	883	950	1000
6.5	154	200	262	308	354	415	462	508	569	616	662	723	770	815	877	924
7	143	186	243	286	329	386	429	471	529	572	614	671	715	757	814	858
7.5	133	173	227	266	307	360	399	440	493	532	573	627	665	707	760	798
8	125	163	213	250	288	318	375	413	463	500	538	588	625	663	713	750
8.5	118	153	200	214	271	318	351	388	435	468	506	553	585	624	672	702
9	111	144	189	222	256	300	333	367	411	444	478	522	555	589	613	666
9.5	105	137	179	210	242	284	315	347	389	420	453	495	525	559	603	630
10	100	130	170	200	210	270	300	310	370	400	430	470	500	530	570	600
10.5								314	352	381	410	448	476	505	543	571
11							300	336	367	390	427	455	482	518	545	
11.5								322	348	374	409	435	461	496	522	
12								308	333	358	392	417	442	475	500	
12.5									320	344	376	400	424	456	480	
13									308	331	362	385	408	418	462	
13.5										319	348	370	393	422	444	
14										307	336	357	379	407	429	
14.5											324	345	363	393	414	
15											311	333	353	380	400	
15.5											303	323	342	368	387	
16												313	332	356	375	
16.5												303	321	345	364	
17													312	335	353	
17.5													302	326	343	
18														317	333	
18.5														308	324	
19														300	316	
19.5															308	

1) Estimate height of target and locate across the top.
2) Measure height of target in mils and locate down the side.
3) Move down from the top and right from the side to find the range.

$$\frac{\text{Height of target (yards)} \times 1000}{\text{Height of target (mils)}} = \text{Range}$$

Figure E-4. Range estimation tables.

TC 23-14

APPENDIX F
SNIPER PATROL ORDERS

1. SITUATION (This describes the enemy and friendly situation as it applies to your patrol.)

   a. Enemy Forces:

   (1) Weather: This is described by using the following format:

   (a) Past: Weather in past 48 hours.

   (b) Present: Current weather.

   (c) Predicted: Weather predicted in next 48 hours or during your mission.

   (d) Effects: How weather will affect both your capabilities and the enemy's.

The following data will be noted using the format below:

DATE	BMNT	SUN RISE	SUN SET	EENT	MOON RISE	MOON SET	PHASE	TEMP (HI/LO)	WIND (DIR/VEL)	HUM %

   (2) Terrain: Explain the effects of the terrain on your capabilities and the enemy's using the key word OCOKA:

   (a) Observation and fields of fire -

   (b) Cover and concealment -

   (c) Obstacles -

   (d) Key terrain -

   (e) Avenues of approach -

   (3) Enemy identification: Information on enemy uniforms, equipment, weapons, and type of unit (for example, the enemy is the 14th Motorized Rifle Battalion; they are armed with AK-47s, RPKs, and SVDs; they are wearing standard camouflage uniforms; and transportation is by BTR-60PB).

F-1

(4) Enemy location: Where they are currently located, and whether they are patrolling in other areas.

(5) Enemy activities: What they are currently doing.

(6) Enemy strength: Their strength in numbers or percent, and the sizes of units in specific locations.

(7) Enemy probable courses of action: What they are expected to do within the next 24 to 48 hours.

b. Friendly Forces:

(1) Mission of the next higher unit: What is the mission of the next higher unit to which the sniper team(s) is attached?

(2) Location and planned actions of units on the left/right/front/rear: Provide any information on units that might be nearby or operating in your area of operation while on your mission.

(3) Units providing fire support: Any unit providing fire support to include artillery, mortar, naval gunfire, aircraft, and helicopter gunships.

c. Attachments and Detachments: Used if any additional personnel are attached to the sniper team during the mission, or when several sniper teams are on the same mission.

2. MISSION (A brief and concise statement of what you are to accomplish. It contains the five W's [what, where, when, why, and who].)

3. EXECUTION

a. Concept of the Operation:

(1) Scheme of maneuver: A brief paragraph that explains how you plan to accomplish the mission from time of departure to time of return and debrief.

(2) Fire support: List all the sniper team's on-call targets and additional targets.

UNIT	CALL SIGN	FREQ	TYPE SPT	TGT NO	GRID	REMARKS

b. Mission of Subordinate Elements:

   (1) Teams: Used if more than one team is on the same mission. Explain specific missions of teams.

   (2) Special teams/key individuals: Special tasks assigned to certain team members during the course of the mission.

c. Coordinating Instructions:

   (1) Time of departure/time of return/time of debrief: Record these times.

   (2) Departure/reentry of friendly front lines:

      (a) Departure: With whom will you coordinate, what unit, time and place for passage of FFL, location of friendly positions, frequencies, call signs, passwords, pyro signals, fire support, and guides. Gain any new information on the enemy from the front line unit. Also, plan in case of enemy contact after or during passage of FFL.

      (b) Reentry: With whom will you coordinate, what unit, time and place for reentry, near and far recognition signals, reentry rally point (grid), plan if under enemy pressure.

   (3) Routes, primary/alternate: Use the following format to explain:

   Route, primary:

FROM (GRID)	TO (GRID)	MAG AZ	DISTANCE	CKPT NO

   Route, alternate:

FROM (GRID)	TO (GRID)	MAG AZ	DISTANCE	CKPT NO

   (4) Rally points and actions at rally points: Explain where you are going to designate each type of rally point (grid) and your actions at each rally point.

      (a) Initial rally point (inside friendly lines).

(b) En route rally points. (How long will you wait there, how will they be selected?)

(c) Objective rally point.

(d) Link-up rally point (if linkup is used).

(e) Reentry rally point (in close proximity of FFL).

(5) Movement techniques: Type of movement techniques used (Who is lead man and what are his responsibilities; who is the rear man and what are his responsibilities?).

(6) Actions at the objective:

(a) Operating with a security patrol. Explain where and when sniper team links up with the security patrol, what unit they are from, call sign, and frequencies. Explain how the security patrol will support the sniper team (hide construction, reaction/rescue force, and so forth) and when the sniper team will separate from the security patrol (security patrol leader is in charge of the team until sniper team separates).

(b) Actions at the ORP. Give location of ORP (grid), team's actions in ORP, AZ and distance from ORP to T-FFP/FFP, movement techniques, and actions on enemy contact.

(c) Actions at T-FFP. Explain location (grid), team's actions at the T-FFP, how you will recon and secure it, actions on enemy contact, and what to do if there is no suitable FFP.

(d) Movement into final firing position. Explain how you plan to occupy and secure it. Explain how you will conduct the initial hasty search, and what is the code word used to inform higher headquarters that you are in position. Also, explain actions taken upon enemy contact.

(e) Concealed position construction. Describe type of position to be used, how and what is needed to construct it, who digs, who pulls security, how long the shifts will be, where the spoil is to be put, and how to use the security patrol if they are to assist in construction of the position.

F-4

(f) Actions in position. Once construction is complete, explain how you are going to place and store equipment. (Draw a diagram depicting this.) Explain how you are going to assign duties. You must develop a work, observe, rest, and alert plan inside the hide position (who observes, who rests, how long).

DATE(S)	ACTION	TIME	NAME	REMARKS

(g) Actions on enemy contact in the hide position. Explain actions if discovered by the enemy (designate escape routes with AZ and distances to rally points). Tell who goes first and who covers.

(h) FFP departure. Explain how you will exit (who goes first, who covers). Give general AZ, distance, and time you plan to exit.

(i) Special instructions. Explain any additional information on the patrol not covered in the above paragraphs.

(7) Actions on enemy contact: Explain what sniper immediate action drills you will use when faced with the following:

(a) Air attack.

(b) Chance contact.

(c) Ambush.

(d) Indirect fire.

(e) Sniper fire.

(f) Minefield/booby traps.

(8) Action at halts: When will you conduct security halts, how will the team be positioned, and what are your actions?

(9) Actions at danger areas: List possible danger areas using the following format:

LOCATION (should be in order of movement	TYPE DANGER AREA

Explain movement techniques used to cross roads, open areas, and streams, and actions upon enemy contact while crossing.

(10) Fire support:

(11) Rehearsals: Where and when will they take place, and what is the uniform and equipment?

(12) Debriefing: When, where, and with whom. Turn in all logbooks, field sketches, and notes.

(13) Annexes: When truck or aircraft support is being used, or for linkup procedures.

4. SERVICE AND SUPPORT

   a. Rations: Date and time of pickup of Class 1, how much will be carried, and times food will be eaten during mission.

   b. Arms and Ammunition: Explain who will draw what from where and how much, where it will be carried during mission, and date and time for test fire and zero.

   c. Special Equipment: Determine what special equipment is needed, who will draw it, from where, and who will carry it.

   d. Uniform and Equipment Common to All: List uniform and equipment that will be the same for each member, how it will be worn and when (if changed since warning order).

   e. Dead and Wounded Friendly Personnel: Procedures to be used for dead and wounded friendly personnel.

5. COMMAND AND SIGNAL

   a. Command: Who is in command of team, where is his position during movement?

   b. Signal:

(1) Frequencies and call signs: Use the following format:

UNIT	FREQ	CALL SIGN

(2) Pyrotechnic signals: Explain any pyro signals used.

(3) Arm-and-hand signals: Explain the seven arm-and-hand signals used by sniper teams.

(4) Challenge and password: Use the following format:

DATE	CHALLENGE	PASSWORD	RUNNING PASSWORD

TC 23-14

APPENDIX G
REPRODUCIBLE FORMS

This appendix provides blank copies of four DA Forms; DA Form 5785-R (Sniper's Data Card), DA Form 5786-R (Sniper's Observation Log), DA Form 5787-R (Sniper's Range Card), and DA Form 5788-R (Military Sketch). These forms are not available through normal supply channels. They may be reproduced locally on 5x7 paper.

SNIPER'S DATA CARD

For use of this form, see TC 23-14; the proponent agency is TRADOC

DISTANCE TO TARGET _____ METERS

RANGE	RIFLE AND SCOPE NO		DATE	ELEVATION		WINDAGE	
				USED	CORRECT	USED	CORRECT
AMMO	LIGHT	MIRAGE	TEMP	HOUR	HOLD		

LIGHT (clock diagram: 12, 3, 6, 9)

WIND VELOCITY / DIRECTION (clock diagram: 12, 3, 6, 9)

SHOT	1	2	3	4	5	6	7	8	9	10	REMARKS
ELEV											
WIND											
C											
A											
L											
L											

NOTE: THE REQUIRED TARGETS WILL BE DRAWN IN BY HAND TO MEET THE NEEDS OF THE UNIT.

DA FORM 5785-R, JUN 89

SNIPER'S OBSERVATION LOG

For use of this form, see TC 23-14; the proponent agency is TRADOC

SHEET _____ OF _____ SHEETS

ORIGINATOR:

DATE/TIME:

LOCATION:

SERIAL	TIME	GRID COORDINATE	EVENT	ACTIONS OR REMARKS

DA FORM 5786-R, JUN 89

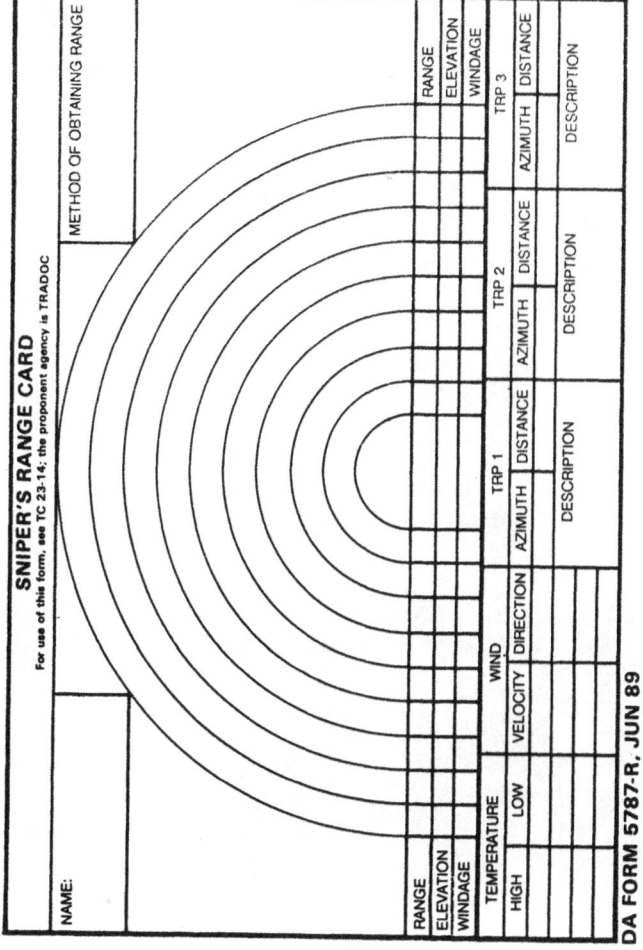

MILITARY SKETCH

For use of this form, see TC 23-14; the proponent agency is TRADOC

REMARKS:

REMARKS:

SKETCH NAME: _____
GRID COORDINATE: _____
WEATHER: _____

MAGNETIC AZIMUTH

SKETCH # _____
OF _____
SCALE _____

NAME: _____
RANK: _____ DATE/TIME: _____

DA FORM 5788-R, JUN 89

GLOSSARY

ALICE	all-purpose, lightweight carrying equipment
ART	auto-ranging telescope
ATTN	attention
AZ	azimuth
BMNT	beginning morning nautical twilight
CKPT	checkpoint
CLP	cleaner, lubricant, preservative
CP	command post
CQ	charge of quarters
DA	Department of the Army
DIR	direction
EDRE	emergency deployment readiness exercise
EENT	end evening nautical twilight
FFL	friendly front lines
FFP	final firing position
FM	field manual, frequency modulation
FRAGO	fragmentary order
FREQ	frequency
FSK	frequency shift keyed
HQ	headquarters
IAW	in accordance with
KIM	keep in memory
LSA	lubricating oil, weapons, semifluid
MAG	magnetic
mm	millimeter
MOA	minute of angle
N	north
NCO	noncommissioned officer
NE	northeast
NO	number
NOD	night observation device
OP	observation post
OPFOR	opposing forces
ORP	objective rally point

RBC	rifle bore cleaner
S	south
SEO	sniper employment officer
SPT	support
SW	southwest
T-FFP	tentative final firing point
TGT	target
TM	technical manual
TRADOC	Training and Doctrine Command
TRP	target reference point
TSFO	training set fire observation
US	United States
VEL	velocity
VHF	very high frequency

TC 23-14

REFERENCES

REQUIRED PUBLICATIONS

Required publications are sources that users must read in order to understand or to comply with this publication.

Field Manuals (FMs)

5-20	Camouflage
7-7	The Mechanized Infantry Platoon and Squad (APC)
7-7J	The Mechanized Infantry Platoon and Squad (Bradley)
7-8 (HTF)	The Infantry Platoon and Squad (Infantry, Airborne, Air Assault, Ranger) (How to Fight)
7-10 (HTF)	The Infantry Rifle Company (Infantry, Airborne, Air Assault, Ranger) (How to Fight)
21-26	Map Reading and Land Navigation
21-75	Combat Skills of the Soldier
23-9	M16A1 Rifle and Rifle Marksmanship
25-100	Training the Force
90-10 (HTF)	Military Operations on Urbanized Terrain (MOUT) (How to Fight)
90-10-1 (HTF)	An Infantryman's Guide to Urban Combat (How to Fight)

Soldier's Training Publications (STPs)

21-1-SMCT	Soldier's Manual of Common Tasks (Skill Level 1)
7-11BCHM14-SM-TG	Soldier's Manual and Trainer's Guide MOS 11B, 11C, 11H, 11M Infantry Skill Levels 1/2/3/4

Department of the Army Forms (DA Forms)

5785-R	Sniper's Data Card
5786-R	Sniper's Observation Log
5787-R	Sniper's Range Card
5788-R	Military Sketch

RELATED PUBLICATIONS

Related publications are sources of additional information. They are not required in order to understand this publication.

Field Manuals (FMs)

6-30	Observed Fire Procedures
7-20	The Infantry Battalion (Infantry, Airborne, and Air Assault)
7-30	Infantry, Airborne, and Air Assault Brigade Operations
21-31	Topographic Symbols
21-76	Survival
24-1	Combat Communications
25-4	How to Conduct Training Exercises
25-7	Training Ranges
26-2	Management of Stress in Army Operations
34-1	Intelligence and Electronic Warfare Operations
34-60	Counterintelligence
71-2	The Tank and Mechanized Infantry Battalion Task Force
90-3 (HTF)	Desert Operations (How to Fight)

90-5 (HTF)	Jungle Operations (How to Fight)
90-6	Mountain Operations
90-8	Counterguerilla Operations

Technical Manuals (TMs)

5-200	Camouflage Materials
9-1005-223-20	Organizational Maintenance Manual (Including Repair Parts and Special Tools List) for Rifle 7.62-mm, M14 W/E (FSN 1005-589-1271), M14A1 W/E (1005-072-5011) and Bipod Rifle, M2 (1005-711-6202)
9-1005-306-10	Operator's Manual, M24 Sniper Weapon System (SWS)
9-1240-381-10	Operator's Manual: Binocular M19 W/E (NSN 1240-00-930-3833)
9-6650-212-12	Operator's and Organizational Maintenance Manual: Telescope, Observation M49, W/E (FSN 6650-530-0960)
9-6920-210-14	Operator's, Organizational, Direct Support and General Support Maintenance Manual (Including Basic Issue Items List and Repair Parts List) for Small Arms Targets and Target Material
11-5820-667-12	Operator's and Organizational Maintenance Manual: Radio Set, AN/PRC-77 (NSN 5820-00-930-3724) (Including Receiver-Transmitter, Radio, RT 841/PRC-77) (5820-00-930-3725)
11-5855-213-10	Operator's Manual for Night Vision Sight, Individual Served Weapon, AN/PVS-4
11-5855-262-10-1	(NSN 5855-01-228-0939) Litton Model M972/M973 (NSN 5855-01-107-5925)
11-5860-201-10	Operator's Manual: Laser Infrared Observation Set, AN/GVS-5
43-0001-14	(C) Surveillance, Target Acquisition, Night Observation (STANO) Equipment and Systems (U)

Training Circulars (TCs)

23-11   Starlight Scope Small Hand-held or
                 Individual Weapons Mounted, Model
                 NO. 6060

23-18   Night Observation Device, Medium Range
                 (NODMR)

INDEX

aiming, 3-3

ammunition
  blanks, 2-27
  special ball, 2-26

binoculars
  M19, 2-28, 2-29 (illus)
  M22, 2-36 (illus)

breath control, 3-7

call for fire, B-10

camouflage
  artificial, 4-3
  blending, 4-2
  deceiving, 4-2
  equipment, 4-5
  face paint, 4-3
  field expedient, 4-4
  ghillie suit, 2-23, 4-3, 4-4 (illus)
  hiding, 4-2
  natural, 4-3

clothing (camouflaged)
  boots, 2-23
  ghillie suit, 2-23, 4-3
  hat, 2-23

concealment, A-5 thru A-8, B-7, B-9

data card, C-1 thru C-3 (illus), G-4 (R-form)

EDRE, B-11 thru B-14

employment, 1-4
  countersniper, 5-4 thru 5-7
  defensive, 5-3
  MOUT, 5-4
  offensive, 5-2
  retrograde, 5-3
  teams, 5-1
  techniques of movement, 4-9 thru 4-17

firing one round, 3-17

Index-1

firing position, 3-1, 3-2 (illus), 3-3
  belly-hide, 4-23, 4-24 (illus)
  construction, 4-20
  expedient, 4-22, 4-23 (illus)
  hasty, 4-21
  occupation, 4-18, 4-19
  selection, 4-17, B-1
  semipermanent hide, 4-25, 4-26 (illus)
  urban terrain, 4-27

holding off, 3-18, 3-19, E-3 (illus)

information, recording, 4-32 thru 4-40, <u>see also</u> data card, military sketches, observation log, and range cards

KIM games, A-10

land navigation exercise, A-9, A-10, B-11

laser observation set, AN/GVS-5, 2-35 (illus)

M14 National Match rifle, 2-1,
    assembly, 2-3
    bolt alignment, 2-3
    bolt assembly, 2-3
    care and maintenance, 2-3
    disassembly, 2-2
    inspection, 2-2
    rear sights, 2-4
    trigger assembly, 2-4

M24 sniper rifle, 2-6
    bolt alignment, 2-7
    bolt assembly, 2-7
    care and maintenance, 2-9
    disassembly, 2-10, 2-11 (illus)
    inspection, 2-8
    safety, 2-6 (illus)
    sights, 2-12
    trigger assembly, 2-8

marksmanship fundamentals, Chap 3, B-5

measurements
    mils, D-1, D-2 (illus)
    minute of angle, D-1, D-2 (illus)

mission, 1-3

movement
  route selection, B-1
  rules, 4-8
  techniques, 4-9 thru 4-17, B-2

night vision goggles
  AN/PVS-5, 2-31, 2-32 (illus)
  AN/PVS-7A, 2-35, 2-36 (illus)

night vision sight, AN/PVS-4, 2-30 (illus)

observation
  detailed search, 4-28, 4-29 (illus)
  hasty search, 4-28
  logbook, 4-39, 4-40 (illus), G-4 (R-form)
  techniques, B-3

observation log, 4-39, 4-40 (illus), B-4, G-5 (R-form)

orders, example, F-1 thru F-7

organization, 1-4
  SEO, 1-4
  team leader, 1-5
  teams, 1-4, 1-5

qualifications, 1-5 thru 1-7

radios
  AN/PRC-77, 2-33 (illus)
  AN/PRC-119, 2-34 (illus)

range
  card, 4-32, 4-33 (illus), 4-34 (illus), 4-42, G-2 (R-form)
  estimation, 4-40 thru 4-44

range cards, 4-32, 4-33 (illus), 4-34 (illus), A-4, B-3, G-2 (R-form)

range estimation methods
  100-meter-unit-of-measure, 4-41
  appearance-of-object, 4-42
  ART I or II scopes, 4-42
  bracketing, 4-42
  combination, 4-44
  exercises, A-8
  mil-relation formula, 4-42, 4-43 (illus)
  paper strip, 4-40
  range card, 4-42

tables, E-5
task, B-3

rifles, see also M14 and M24,
  care of, 2-3, 2-9
  inspection, 2-2, 2-8

rifle systems
  M21, 2-1 thru 2-4
  M24, 2-5 (illus) thru 2-15

shot calling
  iron sights, 3-10
  telescopic sights, 3-11

sights
  alignment, 3-4, 3-5 (illus)
  iron, 2-12 thru 2-14
  National Match, 2-4 (illus)
  picture, 3-5, 3-6 (illus)

sketches, military, 4-34, 4-35 (illus), 4-36 (illus), 4-37
  (illus), 4-38, 4-39 (illus), B-4, G-3 (R-form)

sniperscopes
  auto-ranging telescope, 2-15, 2-16 (illus)
  Leupold M3A telescope, 2-22
  magnification, 2-17, 2-22
  mount, 2-18, 2-24
  zeroing, 2-20, 2-25

sustainment program, B-1 thru B-14

target detection, A-8, A-9 (illus), B-3, B-10

target indicators
  disturbance of wildlife, 4-2
  improper camouflage, 4-2
  movement, 4-1
  odors, 4-2
  sound, 4-1

targets, moving
  leading, 3-20, 3-21
  tracking, 3-20
  trapping, 3-20

target selection
  factors, 4-30

key targets, 4-31
task, B-3

telescope, observation, 2-27, 2-28 (illus)

Threat, identify, B-3

training notes
  field fire, A-4, A-5
  practice fire, A-4
  ranges, A-1, A-2
  targets, A-2, A-3 (illus)
  zeroing, A-4

trajectory chart, E-4

trigger control, 3-8, 3-9

weather effects
  humidity, 3-17
  light, 3-17
  mirages, 3-15 (illus)
  temperature, 3-16 (illus)
  wind, 3-12, 3-13, 3-14, E-2, E-3

zeroing, 2-20, 2-25, 3-11, A-4

www.ingramcontent.com/pod-product-compliance
Lightning Source LLC
LaVergne TN
LVHW041619070426
835507LV00008B/342